CONSTRUCTION
COST ESTIMATING
FOR PROJECT
CONTROL

CONSTRUCTION COST ESTIMATING FOR PROJECT CONTROL

James M. Neil

Morrison-Knudsen Co., Inc.

Formerly Associate Professor of Civil Engineering
Texas A&M University

Prentice-Hall, Inc., Englewood Cliffs, NJ 07632

Library of Congress Cataloging in Publication Data

Neil, James M.
    Construction cost estimating for project control.

    Includes index.
    1. Building—Estimates.   2. Construction industry—
Management.   I. Title.
TH435.N34   624.1′068′1   81-5211
ISBN 0-13-168757-3   AACR2

Editorial/production supervision and
    interior design by Karen Skrable
Manufacturing buyer: Joyce Levatino
Cover design by Diane Saxe

Printed in the United States of America

10   9   8   7   6   5   4   3   2   1

Prentice-Hall International, Inc., *London*
Prentice-Hall of Australia Pty. Limited, *Sydney*
Prentice-Hall of Canada, Ltd., *Toronto*
Prentice-Hall of India Private Limited, *New Delhi*
Prentice-Hall of Japan, Inc., *Tokyo*
Prentice-Hall of Southeast Asia Pte. Ltd., *Singapore*
Whitehall Books Limited, *Wellington, New Zealand*

CONTENTS

19 APPROXIMATE ESTIMATING *312*

PREFACE

Dun & Bradstreet reported the business failure of 154 construction contractors in November of 1976, up 12% from the year before. During the first 11 months of 1976 a total of 2280 contractors experienced business failure, up 7% from the previous period. Certainly, the recession of the prior 2 years was a factor since the construction industry was one of the heaviest hit and the slowest to begin recovery. In fact, one out of every five business failures in the mid-1970s was a construction contractor. In some areas 20 out of every 100 contractors went bankrupt, a rate much higher than that experienced in periods of normal construction activity when one out of 100 is the failure rate. But, even in the difficult mid-1970s, more contractors remained in business than dropped out which would indicate that there is a way to be successful as a contractor in both good and bad years. Why are these contractors still in business? What is it that they do which sets them apart from those that fail? While this text does not propose to provide the complete answer to either of these questions it does pursue the theory that most contractor failures can be traced to faulty cost engineering.

Cost engineering is a relatively new term in the world of the contractor. Few engineers would think of it as a function for which they might be responsible. What then is cost engineering? The American Association of Cost Engineers defines it as

Cost Engineering: *That area of engineering practice where engineering judgement and experience are utilized in the application of scientific principles and techniques to problems of cost estimation, cost control, business planning, and management science.*

In light of this definition, it is obvious that cost engineering must be understood and practiced by every contractor since cost engineering is concerned with everything that affects the ultimate goal of every businessman—the making of a reasonable profit. For a contractor to go bankrupt there must be a failure in the cost engineering area. It could be faulty cost estimation, sloppy cost control, poor business planning, unprofessional management, or a combination of these.

This text has been written to provide a basic introduction to cost engineering of construction through detailed presentation of cost estimation and its relationship to the other project control functions of scheduling, budgeting, and cost control. This book is designed for use as a text for a college or university course or for use in an in-house training program for industry personnel who are first undertaking job responsibilities in the construction project control area.

While the book concentrates on cost estimating, its approach to the subject gives considerable coverage to the contracting function, organization for construction, the total cost-engineering process for construction, and methods of construction. It is designed for education of both design and construction engineers and targets on the following objectives

1. To identify and examine the many components of construction cost and their variability.
2. To present a system for estimate development for major construction.
3. To emphasize that cost estimation is not an isolated function whose end is the preparation of a bid. Estimating must be organized to support and facilitate the entire project control process.

CONSTRUCTION COST ESTIMATING FOR PROJECT CONTROL

ESTIMATING CASE STUDY: THE MONTREAL OLYMPICS COMPLEX

1

INTRODUCTION

Construction is inherently risky because a project must be priced before it is produced. Clients need prices during the planning and conceptual stages to permit development of budgets and financing plans. Then, as the designs are fleshed out, there is a need for updates on prices so that adjustments, if necessary, can be made to budgets or scope of work. Price is all important to a contractor bidding a fixed price contract; the organization must live or die with the bid price. Unfortunately, the accuracy record for estimating construction costs is poor; all too often key factors which affect cost are overlooked or undervalued. To emphasize and illustrate this, this text begins with a case study.

The subject of this study is the Olympics Complex in Montreal, Canada, built to accommodate the 1976 Summer Olympics. This complex is a classic among the many projects each year which drive clients and contractors into tears and bankruptcy because of cost overruns that develop between the budget phase and completion. Is the problem inflation? Certainly, inflation has an effect. Mostly, however, the problem is one of poor planning, design, estimating, and execution, each error contributing its share to the final financial nightmare. Hopefully, by reviewing this project and thoroughly studying following chapters you will become aware of those things that ultimately cost money to a contractor and client and will not similarly be caught short as a contractor or member of a contractor or client's staff when preparing a cost estimate.

BACKGROUND

On July 17, 1976, a young man and a young woman sprinted into the main stadium of the XXI Olympiad in Montreal carrying the traditional flaming torch. Of the millions who watched the event in person or on TV, few realized that these runners were entering a still incomplete stadium, a stadium that had been in planning since 1970. Until almost the last moment, the site had been the scene of frenzied activity as essential construction tasks were completed, temporary structures were erected to mask the incomplete portions, and all evidence of construction was evacuated from the area.

The main stadium was but one of many facilities that were part of the total. A velodrome for bicycle events, swimming pools, boat basins, an equestrian center, roads, walks, subways, practice fields, and many other structures plus landscaping comprised the total project. But the main stadium was by far the largest and most costly structure. Estimated by Montreal's mayor to cost $40 million in 1970, out of a total estimated complex cost of $120 million, the main stadium eventually cost in excess of $836 million and the total price exceeded $1.5 billion. What went wrong?

THE PROBLEMS

Design

The structures were ultramodern and dramatic in concept; they required new and complex construction techniques for the contractors who would execute the designs. The main stadium was designed to look from the top much like an elliptical seashell with a handle. The seashell portion has an elliptical opening in the center that, under the original design, could be open or covered with a retractable fabric cover controlled from a mast that partially overhangs the opening. As a cost-cutting measure, both the mast and the cover were eventually deleted. From the side, the plane of the opening at the top rises slightly from the mast (handle) end of the structure to the opposite end. (See Fig. 1-1)

The interior of the stadium was designed to provide pillar-free viewing by all spectators. To accomplish this, the main structural mem-

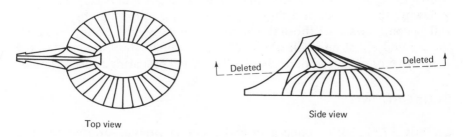

Top view   Side view

Fig. 1-1   Main Olympic Stadium.

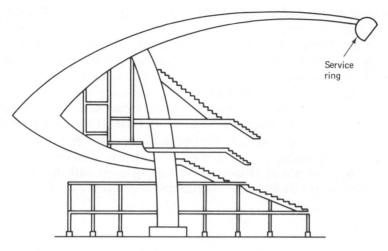

Fig. 1-2 Typical rib.

bers were designed as complex ribs as shown in Fig. 1-2. The ribs were assembled in the field by gluing the parts together with epoxy and post-tensioning a number of precast concrete components. The ribs terminated at the top on a section of a hollow ring that forms the perimeter of the ellipitical opening. This ring houses lighting and other support systems. Because of the sloping top design of the stadium, the ribs were not of the same size. As might be expected, it was an impossible task to assemble and align them perfectly so misalignments approaching 6 in. were encountered. Such problems should have been expected and flexibility incorporated in the design. However, in this case, the design required almost exact alignment of adjoining ribs to permit the threading of posttensioning cable through tubes in the ring.

The posttensioned design feature added further headaches and costs during the winter months when water got into the many tubes and froze before cable had been threaded. Removal of the ice required expensive drilling and contributed further to the loss of time.

Still another problem attributable to a design that did not consider constructibility arose when it was determined that no scaffolding could be used inside the stadium to support the ribs and workers since scaffolding would unduly restrict access into the stadium area. The solution was to use many cranes, some to hold the ribs into position and others to hoist workers, tools, and materials to the overhead work positions.

Another example of cost increases attributable to complex design involved the viaduct on the main road passing the main stadium. The viaduct was on a straight section of roadway and could have been designed rather simply. However, the edge of the roadway was designed as a walk with a series of overhanging parapets with rounded exterior areas from which pedestrians could view the Olympic site. The support for the viaduct's main span looked much like the legs of an inverted swivel chair, being formed of four outreaching concrete legs each requiring numerous prestressing cables for strength. Formwork for some sections of the viaduct was reported to cost as much as $400 per square yard, about 15 times the cost of routine commercial formwork at the time. Overall, the viaduct, which could have cost as little as $5 million using a conventional design cost $14 million or $180 per square foot. Indicative of the complexity of the design was the fact that no contractor would take the project on a fixed price contract. The contractor who built it did so on a cost-plus-fixed-fee arrangement and then only with the condition that he was not responsible for the completed structure.

Labor

Labor was union. Approximately 80 days were lost to strikes and the equivalent of about another 20 days were lost through slowdowns. The project had all the qualities which tempt labor unions to take advantage of the client—there was a fixed deadline, labor was scarce, and there were no agreements between labor and management to restrict strikes or other union activity during the course of the project. There was some sabotage and it was eventually necessary for the client to use police to control labor entry into the site so that known troublemakers could be denied access. The situation was so bad that, in early 1976, the Province of Quebec issued an ultimatum to the workers telling them to either get on with the work or the project would be closed down and the Olympics moved to other facilities. It is interesting to note that production at the main stadium during March and April 1976, following the ultimatum, was twice that recorded during the previous 6 months. While better weather conditions and greater experience by the contractors contributed to a 500% improvement, the ultimatum has to be counted as significant. Not to be outdone, the plumbers and electricians went into a slowdown during the last weeks of the project forcing a delay in the turnover of the project to the Olympic Organizing Committee from June 6 to June 14.

New Construction Technique

The use of the epoxy-glued, posttensioned structural members noted earlier was completely new to North American contractors although the technique has been used successfully in Europe. With any new technique there is a learning process which can be slow, painful, and expensive and this proved to be the case in Montreal.

Resource Shortages

The heavy concentration of construction activity in the Montreal area in support of the Olympics construction literally exhausted all local sources of labor, materials, and equipment. Consequently, these had to be imported from other areas of Canada and the United States. When labor is in short supply and the client is operating under a union agreement, union card-carrying workers must be sought from other areas by the hiring halls since union training programs are so structured that they are not responsive to sudden demands for qualified craftsmen using local personnel. These outside workers (travelers) are not necessarily of high quality and can include troublemakers intentionally introduced into the project.

In the case of construction equipment and materials, those obtained from outside areas carried with them the premium cost of transportation to Montreal often plus a higher purchase or lease price because of the circumstances.

Weather

Many construction operations cannot be performed in the Montreal area during the colder months unless protective measures are employed. Such protective measures cost added dollars, yet are essential on a tight-deadline contract such as this. The cost of heating measures at Montreal was about $400,000 per day at their peak.

Scheduling

A critical path network was prepared for construction of major Olympic facilities. Unfortunately, it had the earmarks of a schedule prepared by theorists unfamiliar with the real world of construction. So many events were scheduled for simultaneous execution that it would have been physically impossible to accommodate all the work forces at once in the area. Consequently, the schedule was abandoned and everything became a daily crash action.

Fixed Deadline

Since missing the opening date for the Olympics could be a national embarrassment for Canada, this project had a very positive no-later-than deadline. Such a deadline is realistic only if the planning, design, and construction activities precede this date in a coordinated manner with plenty of lead time. In this case, it is estimated that the planning started about 2 years too late for routine planning, design, and construction to occur. The City of Montreal was so late in their preparation of contracts that they could not take the time to seek competitive bids so contracts were awarded to selected contractors on a negotiated basis.

Crowded Working Space

The late award of construction contracts and the fixed deadline for the project forced the contractors into schedule compression. The result was doubling of crews, double shifting, and overtime. Unfortunately, productivity does not increase in proportion to such increases in crew assignments when the project is unable to accommodate the numbers of workers and pieces of equipment. At one time 80 cranes were counted inside the main stadium, some as large as 350-ton capacity. It was impossible for all to operate simultaneously so cranes were waiting on each other and the crews were waiting on the cranes. All the while each craftsman was earning an average of $800 per week with overtime and drawing additional money for lost vacation time. It was estimated that the doubling of cranes at the project increased overall productivity by only 25% because of overcrowding.

Client-Engineer-Constructor Organization and Relationship

The original client for the project was the City of Montreal. However, when the project fell seriously behind and financing appeared out of control, the Province of Quebec took over and put the project in the hands of the Olympic Installation Board. They immediately instituted both design and administrative changes. While their takeover did result in improved project performance, the changing of clients in the middle of a project is certain to result in coordination difficulties among the clients, engineers, and contractors.

Basic design and engineering was accomplished in France. The original drawings were sent to Canada with all dimensioning in the metric system. Since North American contractors were unaccustomed to the metric system, all drawings had to be converted to the English system by Canadian engineers. This is but an example of problems associated with great physical distance and poor coordination between engineer and builder. Since it would be extremely unusual for an engineer to prepare designs that are without fault or that do not require change for some reason during the course of a project, it is essential that the engineer be readily available to a project to facilitate the implementation of these changes. Further, it is absolutely essential that a client or a designated project manager oversee the planning, engineering, and construction of a project so that each is compatible with the other.

Inflation

Inflation has been a problem in the last decade with all construction projects. However, cost overruns attributable to inflation should occur only under the following conditions:

1. The project is extended beyond the projected completion date.
2. Actual rates of inflation exceed projected rates.

Inflation has been with us for a long time. Therefore, it is inexcusable for planners not to project their cash flow over the life of the project and include appropriate inflation factors. It is true that inflation rates in North America went into double digits for 2 years during the Arab oil embargo and this certainly had its effect on the Olympic construction program. However, the project was not extended, so the overall effect of inflation should have been small in comparison with the other factors already discussed.

SUMMARY

In the beginning the Montreal Olympics complex was estimated to cost $120 million and was to be paid for entirely through a lottery and sale of commemorative medals and stamps. In the end, the figure was exceeding $1.5 billion, the final product contained less than originally envisioned, and the citizens of Montreal and the Province of Quebec are looking at years of higher excise and property taxes necessary to retire the huge debt. The main stadium, even with the mast and cover deleted, cost over $13,000 per seat. The size of this figure can best be appreciated when compared to the price tag of the New Orleans Superdome which was built for less than $2400 per seat and was completely enclosed.
The lessons to be learned from the Montreal Olympic construction experience are many. It should be your goal to learn from this and other projects so that you, in evaluating any project for cost, will look closely for those features which translate into higher than average costs. Here are questions of the type you should ask. You may think of many more.

1. Is the design particularly unusual so as to require new construction techniques?
2. Is the design such that most components are one-of-a-kind so as to preclude taking advantage of mass-production techniques or gaining the greater productivity rates that come with repetitive work?
3. Does the design consider constructibility?

4. Is the overall management of the project—client, engineer, contractors—well organized for full coordination, cooperation, and responsiveness?

5. Does the project have a no-later-than completion date?

6. Is the project union? If so, have any agreements been signed to preclude adverse union activity?

7. What effect will weather have on performance? Will supplementary heating be required?

8. What is the availability of resources in the area of the project: labor, construction equipment, and materials?

9. What effect will inflation have on the project?

10. To what extent will crash scheduling be needed?

11. Is there adequate working space for normal crew operations?

In the following chapters we will explore the subject of construction cost estimating in considerable detail. Hopefully, through this case study, you now realize that good cost estimating is a complex subject requiring considerable analysis and judgement. It is not simply the extraction of quantities from design drawings and the factoring of these quantities to determine costs and employee hours.

REVIEW

Questions in Review

1. In what way did the following contribute to the cost overrun of the Montreal Olympics complex?
   a. Design
   b. Labor
   c. Fixed deadline
2. Should inflation be a factor in cost overruns?

Discussion Questions

1. If you had been in charge of the construction of the Montreal Olympics complex, what would have been your approach?
2. Using a project in your area which is known to have suffered from cost overruns, identify what you think were the problems.
3. If a project is changed from one 8-hour shift to two 8-hour shifts per workday, will productivity double?

ORGANIZATION FOR CONSTRUCTION

2

INTRODUCTION

Before launching into specific coverage of estimating, several chapters will be devoted to subjects which form a background for the entire cost-estimating process. This chapter covers organizations for construction and its objective is to provide an overview of the functions that must be performed by any construction organization as well as the more common organizational forms for accomplishing these functions.

KEY INDIVIDUALS AND ORGANIZATIONS

Throughout this text terms are used which refer to individuals or groups in the plan–design–build sequence. These terms and their definitions follow.

Architect

An architect is an individual who designs buildings and their associated landscaping. Most architects are capable of only limited structural, electrical, mechanical, and other specialized design so will normally rely on consulting engineers for this work. An architect may monitor construction on behalf of a client but will not supervise it.

Architect-Engineer (A/E)

This term is usually applied to a business firm which employs both architects and engineers and is capable of complete design work. They also may have the capability to perform construction management services (see below).

Client

Client is a commonly used synonym for the owners of the facilities to be constructed. Clients do basic planning and budgeting for the project and, subject to their capabilities, may handle all or portions of the project management, engineering, procurement, and construction. A client will engage architects, engineering firms, and contractors as necessary to accomplish the desired work.

Construction Manager

There are three uses or definitions for this term. First, a construction manager is a person who directs and manages actual construction for a contractor at the project site. A second, and more recent use of the term, refers to individuals or groups who are engaged in a services contract by a client to perform project management or construction management services in their behalf. In this arrangement, the construction management group will have the responsibility of coordinating design and/or construction work, will prepare contracts for award by the client, will oversee construction for the client and perform other services as agreed upon. Third, a construction management firm may engage in a construction contract in the same way as a general contractor. The firm bids on work and then, if awarded the contract, becomes totally responsible for its execution. Normally, it will subcontract all operations except management.

Engineer

This is a broad term which refers to an individual engaged in specialized design or other work associated with design or construction. Design engineers are classified as civil, electrical, mechanical, environmental, etc., depending on their design work specialty. There are also scheduling, estimating, cost, and construction engineers whose origins may have been in any of the basic engineering disciplines but who have specialized in a particular area.

Engineering-Construction Firm

This type of organization is a combination architect-engineer firm and general contractor. They are capable of executing a complete design-build sequence or any portions of it. This capability is found only among the largest firms.

Force-Account Labor

In-house labor hired as individuals by a contractor are in this category. A contractor will maintain force-account labor only when the workload in a given craft specialty justifies regular employment of these craftsmen or when specialty contractors are not available to handle the work.

General Contractor

This type of contractor seeks contracts for the construction of a total facility or a major portion of it. They generally maintain the capabilities to perform certain of the more common construction operations such as site preparation, concrete placement, and carpentry. For specialty work—utilities, roofing, masonry, finishes, etc.—they frequently engage subcontractors, although they retain the responsibility for the satisfactory performance of these subcontracts.

Project Manager

A project manager is an individual charged with the overall coordination of all facets of a construction program—planning, design, procurement, construction, etc. On very large projects executed under the turnkey concept, a full-time project manager will be appointed within the engineering-construction firm. One project manager may handle several smaller projects.

Specialty Contractor

This contractor performs only specialized construction. Plumbing, electrical, painting, and similar work is normally done by specialty contractors.

Subcontractor

A subcontractor is under contract to another contractor, as opposed to a client, to perform a piece of work. There can be several levels of subcontracting. A general contractor under contract with a client may engage subcontractors for portions of the project. These subcontractors, in turn, may engage other subcontractors. One general contractor can be a subcontractor to another general contractor.

CONTRACTUAL ARRANGEMENTS

The definitions of the previous section imply a number of possible contractual arrangements for accomplishing design and construction work. Some of the more common are

Design by A/E; Construction by General Contractor (GC). The client enters into a design contract with an A/E. The design is executed by the A/E with input from required consulting engineers. The A/E prepares contract documents for the client and assists in the award of a contract to a general contractor. That contract is signed by the client and the GC. Thereafter, the A/E monitors the contract. The GC executes the construction utilizing force account labor and appropriate subcontractors. (see Fig. 2-1)

Design by A/E; Contract with CM Group for Construction. As above, except a construction management firm replaces the general contractor and all work would be done by subcontract.

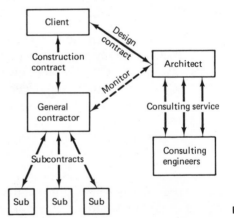

Fig. 2-1

Design by A/E; Service Contract with CM Group for Construction Management. The client enters into a design contract with an A/E firm and a construction management contract with a CM firm. The design, when prepared, is passed to the CM for preparation of contract documents. The CM awards contracts on behalf of the client and thereafter manages the project for the client. (See Fig. 2-2)

Turnkey Contract with Design-Build Firm. A *turnkey* contract is a design-build contract with a single organization. That organization will prepare designs, prepare the budgets for the client, and execute the

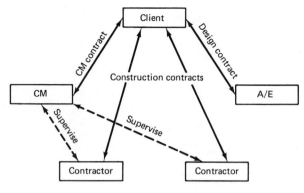

Fig. 2-2

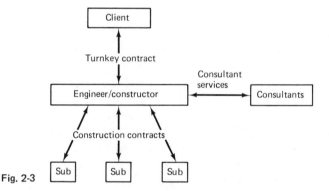

Fig. 2-3

construction using force account and/or subcontracts. Some specialty consultants may be used in the engineering phase. (See Fig. 2-3)

Turnkey Contract with General Contractor. The client engages in a design-build contract with a general contractor. Since GCs do not do design work they contract this to an A/E firm. The GC retains full responsibility for both design and construction and the client looks only to the GC. As usual, the GC will probably utilize a number of subcontractors in doing the construction. (See Fig. 2-4)

Turnkey Contract with A/E Firm. This is similar to a turnkey arrangement with a general contractor. The client contracts with an engineering firm for both design and construction. The A/E designs the project and then contracts with one or more contractors for the work. With this type of contract the A/E is totally responsible for complete design and construction.

Turnkey Contract with CM Firm. For this contracting arrangement the CM firm accepts total design and build responsibility. However, it contracts out both design and construction.

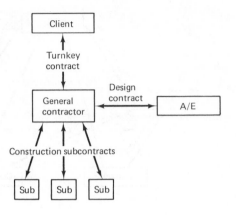

Fig. 2-4

TYPES OF CONSTRUCTION

There is no universal agreement on the types of construction and their inclusive elements. The following groupings are representative.

Residential. This category includes all single-family dwellings and apartments. For some purposes, these are divided into homes, townhouses, low-rise apartments (four stories or less), and high-rise apartments.

Commercial. Stores, office buildings, warehouses, small manufacturing facilities, hospitals, service stations, shopping centers, banks, and nursing homes are typical structures in this category.

Industrial. This category is reserved for major factories, power plants, petrochemical plants, other process plants, and similar industrial installations.

Heavy and Highway. Streets, highways, railroads, airports, subways, bridges, dams, and pipelines are typical projects in this group.

Marine. This grouping includes off-shore platforms, submarine pipelines, structures related to navigation and ocean commerce, and dredging.

Federal Civil Works. Federally funded projects of the types listed previously that are designed to protect population groups or to provide facilities of general benefit to the population are in this group. Flood control structures, locks and dams for navigation, port facilities, and multipurpose reservoirs are typical projects.

Military Construction. These federally funded projects directly support the United States defense effort and include everything from military housing to missile bases.

In setting up its organization, an engineering/construction firm will establish different divisions, each incorporating one or more of the above categories of construction in which it seeks contracts.

FUNCTIONS WITHIN A CONSTRUCTION ORGANIZATION

Large or small, a construction organization will find itself involved in a variety of functions in its day-to-day operations. These may be broadly categorized as management, operating, and special staff functions. For the large firms, separate organizational elements will handle each function within each category. For the smaller firm with only a few professional level employees, the functions must be grouped and each of those employees finds himself or herself with an assortment of responsibilities. It may be necessary for some services to be handled by outside consultants. Unfortunately, none of the functions can be ignored if a contractor is to succeed financially and avoid legal difficulty. Thus, it is incumbent upon all contractors to become familiar with all of them and organize their firms for their most efficient handling.

MANAGEMENT

Leadership of the firm is housed in the management group. They establish the policies, provide the guidance, make the decisions, and structure the organization to meet the established objectives. The results of their abilities are reflected in the balance sheets and the image of the firm as seen by potential clients. It takes much more than a position title to be a manager. It requires knowledge of construction, knowledge of good business and financial practice, an ability to plan and organize; and an ability to attract and mobilize quality employees.

Operating representatives of the management group will be those people designated as project managers, design managers, and construction managers. They, in turn, will have their directions implemented by other supervisors and foremen. Within the management structure there must be clearly established channels of control and authority. Each individual must know to whom he or she is responsible. Further, authority and responsibility must be delegated down the chain of management if the various levels are to be effective and responsive.

OPERATING DIVISIONS

The operating divisions accomplish the business of the organization, that is, perform the actual design and/or construction. An equipment division also may exist as a separate operating division.

Architecture/Engineering Division

This division maintains, to the extent practical, a staff capable of executing any designs for work of the type within the advertised capability of the firm. They will respond to Requests for Proposals (RFP) made by clients interested in selecting a design agency for a proposed project and, if selected for the work, of designing the project. They may utilize outside consultants for specialized work outside their staff capabilities. This group also will handle any special studies or planning work requested by clients. A full engineering division would not be found in a construct-only firm.

Construction Division

The construction division will have the capability to estimate, plan, schedule, cost control, and construct a project as a general contractor, a member of a joint venture, or as a subcontractor. The construction division organizational structure will include the various project staffs. In an engineer-only firm, this division would be replaced by a construction management group if used at all.

Equipment Division

If used, the equipment division is equivalent to a construction equipment rental firm within the organization. They "own" all equipment of the company, maintain it, and "rent" it to the various projects as needed. (See Chap. 10)

SPECIAL STAFF

A special staff section is a support group. They are not the elements that do the work for which the clients pay, but they are equally essential since they maintain the organizational framework within which the operating elements can perform. Those found in a large engineering/construction firm are discussed below.

Marketing

Many clients are extremely selective in their choice of contractors for bidding on their projects. Thus, it is essential that contractors keep abreast of planned construction by potential clients and sell their capabilities to them so that they will be asked to compete. This is done through the Marketing Group.

Personnel Management

This function encompasses recruiting, dismissal, training, and administrative management of the firm's employees. Since employees will make or break a firm it is essential that a firm maintain policies and conditions which attract and retain quality individuals. A high employee turnover rate is a sign of organizational weakness. Either the employees are not receiving adequate financial compensation for their work or the general working environment within the organization is below par. While the personnel office cannot be responsible for all organizational problems affecting employee recruitment and retention, they can do their part by establishing a favorable image with employees at the time of recruitment and thereafter maintain this image through faultless personnel adminisistration. Further, the personnel office, as a special staff office for the management group, will advise management concerning any policies which affect personnel recruitment, retention, and training. They must be particularly aware of such programs as Equal Employment Opportunity (EEO), and the provisions of union contracts so as to best insure their employer's compliance.

Labor Relations

Construction labor will be discussed in detail in Chaps. 6, 7, and 8. In this discussion it will be pointed out that the craft labor will be hired under either open shop (nonunion) or union shop conditions. When operating under union shop conditions, employers must be continually conscious of the union contract(s) under which they operate and be prepared to deal on an almost daily basis with union problems. For large contractors, a separate labor relations staff at both the home office and job level will be used for this purpose. For the small contractors, labor relations become an additional function of management. A labor relations staff may be a subunit of the personnel section.

Legal

All businesses are subject to those laws which govern everyone's daily lives plus a host of laws specifically directed toward commercial business. In any business where contracts are used, these contracts are subject to legal interpretation regardless of the detail with which they are prepared. Claims and lawsuits are commonplace in the construction industry. Small contractors attempt to legally protect themselves through retention of an attorney, hopefully one familiar with construction. Major contractors will maintain a full-time legal staff.

Accounting/Finance

Any business must give attention to the accounting function since accurate financial records are essential for tax purposes and to provide the information needed for future planning. Further, most construction firms rely on financial support from banks and other institutions and these organizations invariably do business only with those construction firms with a satisfactory financial standing proven by a good set of financial records.   Accounting is a special staff function that supports every element of the company in some fashion from payrolls to tax returns.   A very small construction firm may find it uneconomical to retain its own accounting function and will contract for accounting services with a local certified public accountant.   In the larger firms the accounting sections are so large as to require in-house computer systems.   These larger accounting sections also will include experts on business taxation.

Safety

Safety has always been an important subject in construction because accidents result in human suffering and bring financial loss to the victims, their families, and the contractors.   The subject became even more important in 1971 with implementation of the Occupational Safety and Health Act (OSHA).   This act made safety a matter of public law and a strong enforcement organization was established to insure compliance. The Occupational Safety and Health Administration (OSHA also), has adopted volumes of safety requirements previously used as guides by various professional and industrial associations (concensus standards) and written additional safety rules, all of which have the force of law. Through its regional offices, OSHA is empowered with the authority to make scheduled and unscheduled inspections of construction sites and to issue citations for nonconformance which can result in OSHA levied fines or court ordered suspension of operations.   The safety record of an employer also affects the rates paid for Workmen's Compensation Insurance (Chap. 7).   Needless to say, a contractor must give safety constant attention.   For small contractors it is an additional function of all supervisors.   For the larger firms there will be a safety office at the home office plus safety engineers on the larger projects.

Procurement

Procurement includes subcontracting and purchasing.   A subcontract covers services and/or materials involving on-site labor by the subcontractor.   Purchasing covers materials or equipment delivered to the site with installation labor provided by other than the vendor.

Requirements for subcontracts are generated by the engineering or project staff with the procurement personnel handling the contract formalities and master file maintenance.

Purchasing for small contractors normally involves only the routine purchase of materials and equipment used in their contracts, most of which are available off-the-shelf.   Actual purchases are most likely handled by the project personnel and the bookkeeping is handled by whatever accounting section there is. For the major contractor with many large projects, purchasing becomes a major function because of the sheer volume of transactions. This requires extreme formalization of all procurement and warehousing operations if order is to be maintained and projects are to be serviced in a timely manner. In addition to the routine off-the-shelf category of procurement, this section will prepare contract documents for receiving bids and awarding contracts for specification material and equipment items. They will maintain catalogs and data for use by the operating divisions for estimating costs and procurement lead time, handle sale or disposal of excess property, perform expediting and follow-up service on procurement actions, and manage all transportation requirements. All large projects will have purchasing and warehouse staffs at the site to receive, account for, and issue materials and equipment. Also, they will handle certain direct purchase transactions for the construction manager. This project staff is under the direction of the project's construction manager but they will maintain close liaison with the home office procurement section and will provide much of the data used in the procurement reporting system.

Cost Control

One very important function of project management is that of cost control or the management of costs. A large contractor probably will maintain cost control sections within the operating divisions while also maintaining a senior cost control section which serves a coordinating role. Cost control involves the processing of raw information received from the projects, operating divisions, and special staff divisions; relating this information to the various project cost estimates and schedules; and the presentation of results in the form of reports to all levels of company management, the client, and outside agencies. This section will normally have access to a computer for their work. They also will maintain historical data on projects for use by management and the estimating and scheduling sections of the operating divisions. Chapter 5 provides a more detailed discussion of this topic.

Quality Assurance/Quality Control

All construction projects that include structures that can be related to nuclear safety require special procedures which are under the heading of quality assurance/quality control (QA/QC). QA/QC also can be applicable to other types of construction where an extremely high degree of assurance is required that the project meets each and every specification. Currently, QA/QC is receiving its greatest attention in the nuclear power industry but many observers predict that it will soon govern construction in the environmental protection area as well.

To properly understand QA/QC some definitions are in order. *Quality* is an essential property of an item, object, or thing that is used to identify its degree of excellence, grade, or character. *Control* is an action taken to check, test, or otherwise verify quality by obtaining evidence or results. *Assurance* is an action taken to make sure or to establish certainty and confidence. Assume, for example, that QA/QC is to be applied to the placement of concrete. *Quality assurance* is initiated by developing a written procedure for accomplishing this work. This procedure will identify the work to which it applies, list all individuals who have responsibilities connected with the work and identify those responsibilities, describe the way the concrete is to be placed, list quality control destructive and nondestructive examinations required during and after placement, and describe documentation that must be completed to verify that the work was performed as prescribed. Full compliance with this procedure will constitute quality assurance. It must be emphasized that overall QA responsibility rests with top management and that QA inspectors representing top management are separate from the production chain of command. Note that *quality control* is but one component of QA. Concrete is placed and quality controlled by the production personnel under the watchful eye of the QA inspectors. While these actions may be extreme, it is the only way to assure specification conformance in critical construction.

Contractors with projects that include QA/QC as a contractual requirement must establish a separate staff group for this purpose. A control element is organized at main office level with action personnel within each operating group involved in design or construction of QA/QC controlled work.

BUSINESS ORGANIZATIONAL FORMS

Several business organizational forms are open to the firm performing design and/or construction work. Each has its place in the business world and its advantages and disadvantages.

Single Proprietorship

This type of business has one owner who is in direct charge of all activities of the business. There can be any number of employees, both professional and nonprofessional, to handle the various company functions. Such an organization reflects the owner's personality and service to the clients is more personal than with other business forms. This form has a number of disadvantages. The owner must be an all-round expert in the business and be readily available to make decisions at all times. In the event of sickness or absence of the principal, the business may falter; in the event of death, the business is usually terminated. The owner must provide for his or her own fringe benefits. Finally, the owner is totally liable as an individual for all consequences of the business' operations.

Financing of a single proprietorship is by contributed capital or property from the owner or through borrowing. This form of business is taxed through its owner, not as a separate entity; that is, all incomes and losses are incorporated into the income tax return of the owner.

Partnership

There are two or more owners in a partnership. Ownership can be equally split or otherwise as agreed to by the partners. A partnership has the advantages of making more capital available than is the case of single proprietorship and of providing more principals among whom the responsibilities can be divided. In a typical partnership, the partners would specialize; for example, one might handle main office operations while the other handles field operations. Disadvantages for the partnership are several. First, one partner can obligate another. Second, the success of the partnership is highly dependent upon the compatability of the partners. Third, the partnership will be automatically dissolved upon the death of any partner. Fourth, the partners must provide for their own fringe benefits. Fifth, partners share all liability as individuals.

Financing of a partnership is by contributed capital or property from the partners or through borrowing. A partnership is taxed through the partners. All income and losses are passed to the partner in proportion to their ownership in the firm for reporting on their respective income tax returns.

Limited partnerships are those where one or more partners are controlling (general) partners while others are limited.

Corporation

A corporation is a paper individual, liable within its own limits of ability, not those of the individuals who own it. It has several distinctive characteristics:

1. *Separate Legal Entity.* The corporation is separate from its owners. It engages in business as a paper individual, with all rights to acquire and dispose of property, produce goods or services, enter into contracts, and incur liabilities.

2. *Continuity of Life.* The life of the corporation is not related to the life or capacity of its owners. In theory it is perpetual.

3. *Limited Liability.* The corporation, as an individual, is liable for its actions and debts but this liability is not passed to the owners except to the extent that their investments in the corporation would be affected.

4. *Transferability of Ownership.* Ownership in the corporation is obtained through purchase of shares of stock. These shares may be sold to other individuals.

5. *Central Management.* Management of the corporation is through a board of directors selected by the owners (stockholders). This board, in turn, appoints key management personnel who operate the firm on behalf of the stockholders.

6. *Subject to Regulation.* Corporations are chartered by the individual states and must operate in accordance with the laws of those states concerning corporations.

7. *Corporate Taxation.* The corporation is taxed as a unit at the local, state, and federal levels. In addition, any of its profits distributed as dividends to the stockholders will be taxed a second time when reported on the personal income tax returns of these stockholders. Thus, some corporate income is subject to double taxation.

The greatest advantages of a corporation are its limited liability and perpetual life. Additional advantages are its ability to expand, to obtain additional capital through sale of stock, and to provide fringe benefits to all officers and employees through the corporate structure. Its disadvantages lie in double taxation, government regulation, and the loss of personal touch.

Corporations are financed initially through the sale of stock to individuals or groups. This stock is broadly categorized as common and preferred although there are a number of special forms of each. Common stockholders obtain voting rights through purchase of their stock while preferred stockholders do not. Corporations obtain additional funding, as necessary, through sale of bonds or commercial paper, and by borrowing.

Subchapter S Corporation

A Subchapter S (SubS) corporation is a special category of corporation which combines features of a regular corporation and a partnership. SubS is a classification granted by the Internal Revenue Service (IRS) to corporations which meet certain criteria. This classification must be applied for to the IRS—it is not automatic. It can later be withdrawn either at owner request or because the criteria are no longer being met. Basically, a SubS corporation is a corporation with all of the characteristics identified above except for corporate taxation. In a SubS corporation all income and losses flow to the owners (stockholders) and are reported within their individual income tax returns so there is no taxation at the corporate level.

The SubS option is available only if certain criteria are met. They are

1. There can be no more than 10 stockholders at any one time.
2. A nonresident alien may not be a stockholder.
3. Partnerships or trusts may not be stockholders.
4. No more than 80% of the gross receipts can be from sources outside the US.
5. No more than 20% of the gross receipts can be from certain types of passive income such as rents, royalties, interest, dividends, and annuities.
6. Only one class of stock can be issued.

TYPICAL ORGANIZATIONS

As is obvious from previous discussion, many organizational patterns can be established for the engineering/construction firms. Following are several charts that show representative organizations.

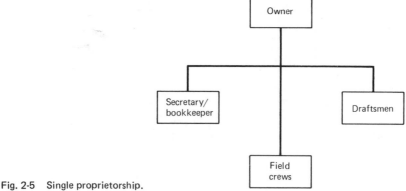

Fig. 2-5   Single proprietorship.

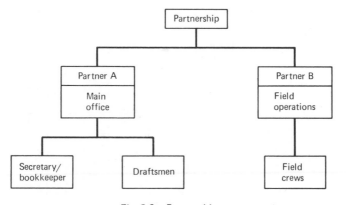

Fig. 2-6   Partnership.

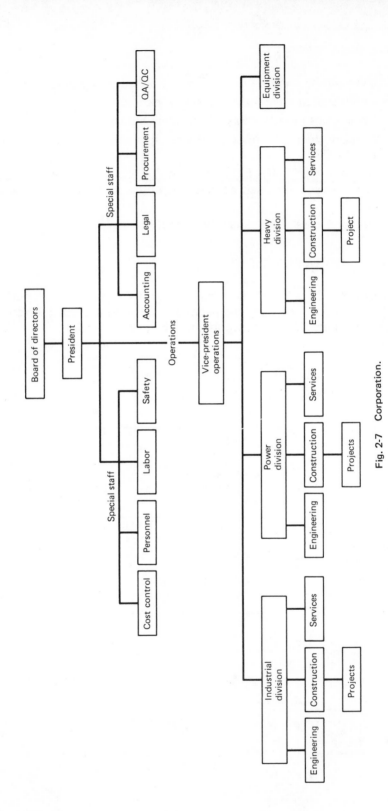

Fig. 2-7 Corporation.

Single Proprietorship

See Fig. 2-5.

Partnership

See Fig. 2-6.

Corporation

See Fig. 2-7.

SUMMARY

The major points to be derived from this chapter are that there are many vital functions to be performed by contractor organizations in addition to the production of designs or the placement of work and there are many ways to organize to handle these functions. There is no single best way of doing things. The type of business, the volume of business, the capabilities and personalities of the principals, tax considerations, nature of the labor force, work locations, and countless other factors enter into the decision. Most company histories will show that they started with one structure and reorganized almost annually as business grew and conditions changed. The main objective for a firm is to find a structure that works for the current situation while maintaining potential for expansion. Then, as conditions change, the organization is continually fine-tuned for maximum operating economy and efficiency while providing quality output.

REVIEW

Exercises

1. Develop an organizational chart for a small three-partner custom home building firm. One partner is an architect, one a business major, and the third a building constructor.
2. Develop an organizational chart for a corporate construct-only firm. The firm has been specializing in highway construction but is expanding into process plant, power plant, and marine construction. It is not yet big enough to afford separate estimating, planning and scheduling, or cost control units for each specialty.

Questions In Review

1. Distinguish between a construction manager and a project manager.
2. What is force-account labor?

3. What type of firm can handle turnkey projects?

4. A client has engaged an A/E to prepare designs for a project and has engaged a CM to project manage the contract. Diagram the contractual and supervisory relationships among the client, CM, A/E, contractors, and their subcontractors.

5. Describe typical special staff sections of a major engineering/construction company.

6. What is OSHA?

7. What is the difference between quality assurance and quality control?

8. What are the taxation differences between partnerships and regular corporations? Between partnerships and SubS corporations?

9. What are the characteristics of a corporation?

Discussion Questions

1. Why would a small family-owned construction business be operated as a single proprietorship rather than as a SubS corporation?

2. A small contractor intends to specialize in the construction and rental of apartment buildings. He and his family want to organize the business as a SubS corporation. Will he be allowed to do so?

CONSTRUCTION FINANCE TOPICS

3

INTRODUCTION

Cost estimating is concerned with money. Resources—labor, materials, equipment, and time—all eventually are converted into dollars of direct and indirect costs in the development of a bid. Then, management must make a decision on the markup of the job level costs so that the costs of doing business are properly included and the firm makes a reasonable profit. Again, everything is converted to money. Throughout these processes, the estimators and managers must have an understanding of true costs and their derivation. In this chapter you will be introduced to a number of basic financial subjects which affect the costing operation directly or indirectly. The objective is not to make you an expert in these subjects, since each one can be the subject of a complete text, but to give you a sufficient view so you can appreciate their impact on cost estimating.

BALANCE SHEETS

A balance sheet is a summary of the financial status of a business at a point in time. It has a standard format which is used and understood throughout the business world. In this section we will describe its composition and its meaning to a contractor and estimator.

Let us assume that you are interested in becoming a contractor. You know, of course, that getting started in business requires capital because there is equipment to buy, salaries to pay, office and utility costs, etc. If you run to the nearest bank and announce your intention to set up this business and ask them for a loan of $100,000, the first question they probably will ask you is, "What is your personal investment in this business?" In other words, no one is going to take all the risk for you. You must have your own money or property to contribute to the business as evidence that you are also taking a risk. Once this is done, the bank may consider that loan. At this point, let us assume that you have been fortunate enough to accumulate $20,000 in savings and are willing to invest this in the business. Also, you plan to convert your garage into office space for your firm. The garage is valued at $5000 in its present condition. Overall, then, you are going to contribute $25,000 in assets to the business. These contributions are called *owner's equity* and this equity is represented at the moment by *assets* of $20,000 in cash and $5000 in real property. The initial balance sheet for your firm looks like this

Balance Sheet, January 1

ASSETS		OWNER'S EQUITY	
Cash	$20,000	Contributed Capital	$25,000
Real Property	5,000		
	$25,000		$25,000

Note that the assets and owner's equity balance in value. With balance sheets, the left hand side must always balance out to equal the right.

The garage, in its present condition, is unsatisfactory as a business office so you spend the month of January converting it into an office. The total cost of the remodeling work is $5000 and you pay for this work from the firm's checking account. On February 1 your balance sheet looks like this

Balance Sheet, February 1

ASSETS		OWNER'S EQUITY	
Cash	$15,000	Contributed Capital	$25,000
Real Property	10,000		
	$25,000		$25,000

Note that the totals have not changed but you have redistributed your assets to reflect the remodeling action.

You are now ready to go into business but you need supplies, tools, and equipment. The cost of the supplies is $4000, the tools $3000, and the construction and office equipment will cost $43,000. Obviously, the $15,000 in your checking account will not cover these items and you realize that you must maintain a reasonable balance in this account at all times to cover purchases and payroll. So, now you go to see your banker. You tell him or her about your business and your plans for its growth. Hopefully, you have done your homework carefully to show exactly how you plan to make a profit so that you can convince the banker that you are worth investing in. Let us assume the banker is convinced, and agrees to loan you $50,000. The money is deposited in your account (see later discussion on bank loans) and over a period of the next few weeks you buy the supplies, tools, and equipment. On March 1 your balance sheet will show

Balance Sheet, March 1

ASSETS		LIABILITIES	
Cash	$15,000	Bank Loan	$50,000
Supplies	4,000	OWNER'S EQUITY	
Tools	3,000		
Equipment	43,000	Contributed Capital	$25,000
Real Property	10,000		
	$75,000		$75,000

This time a new category has been added: *liabilities.* As its name implies, this is money you owe to someone else. It represents other people's money that is working for you but you maintain an obligation to repay it. Also note, that the right hand side still balances the left but now contains the two categories, liabilities and owner's equity. Thus, the balance feature of a balance sheet can be expressed this way

$$\text{Assets} = \text{Liabilities} + \text{Owner's equity}$$

This equation is an absolute rule. Your assets have been financed by liabilities and owner's equity.

Now you are in business and we will assume that you are able to attract all the work you can handle. During the remainder of the year your operations show the following

Gross income from contracts	$200,000
Supply purchases	105,000
Supplies used	100,000
Depreciation	7,000
Wages and salaries (except yours)	30,000
Office expenses	3,000
Other expenses (no salaries)	19,000

Standard accounting practice has procedures for handling all of these transactions in a formalized manner. We will simplify and short-cut. First, determine the net profit

Gross Income		$200,000
Less: Expenses		
Supplies used	$100,000	
Wages and salaries	30,000	
Depreciation	7,000	
Office expenses	3,000	
Other expenses	19,000	
		159,000
Net income before taxes		$ 41,000

What we have done is subtract the cost of doing business from the gross income received. Note that we subtracted only supplies used, not supplies purchased. You purchased $105,000 of supplies but used only $100,000 worth. Remembering that you already had $4000 worth of supplies on hand, you now should have an inventory valued at $9000. Also, note the depreciation. This represents wearout of the equipment and buildings. You did not actually write a check to someone for depreciation but the value of these items at the end of the year is $7000 less than when you started and is a cost of doing business. The purchase price of a depreciable item is not considered a business expense in the year of purchase; instead, you amortize the initial cost through depreciation over its years of use. But, we are not through yet. This is a business and the state and federal government are always interested in sharing in your success so you must pay income taxes. Assuming that this is a single proprietorship, the business expenses will be added to other allowable deductions on your income tax form and the result is

Gross income		$200,000
Less:		
Business expenses	$159,000	
Personal deductions	8,000*	
		167,000
Taxable income		$ 33,000

*Personal deductions include those for dependents, medical expenses, home taxes and mortgage interest, and similar items not part of the business.

The amount of taxes owned on the taxable income of $33,000 will depend upon current federal and state scales. For purposes of this example, assume the total taxation is $10,000 leaving $23,000 net income after taxes. Unfortunately, there will not be a bundle of $23,000 that we can isolate. Recall that the owner's salary was never considered and that with a single proprietorship the business and the owner become one for taxation purposes. Again, let us make an assumption, this time that you have withdrawn $18,000 from the business during the year as your salary. With these figures, we can finally see where the business stands.

Gross income		$200,000
Less: Business expenses	$159,000	
Owner salary	18,000	
Taxes	10,000	
		187,000
Net profit after taxes		$ 13,000

What this last calculation shows us is that $13,000 profit was realized during the year. You have three options for the use of this money. First, you can take it all as a dividend or bonus without further taxation. Second, you can leave it in the business to encourage its growth—such funds are called *retained earnings*. Third, you can elect to take part as a bonus and leave the remainder as retained earnings. For this example, assume the entire amount is left as retained earnings. The final balance sheet for the year will show

Balance Sheet, December 31

ASSETS			LIABILITIES	
Cash[1]		$23,000	Bank Loan[7]	$45,000
Accts Receivable[2]		4,000	Accts Payable[8]	2,000
Supplies[3]		9,000	**OWNER'S EQUITY**	
Tools[4]		3,000		
			Contributed Capital[9]	25,000
Equipment[5]	43,000			
Less: Deprec.	6,500		Retained Earnings[10]	13,000
		36,500		
Real Property[6]	10,000			
Less: Deprec.	500			
		9,500		
		$85,000		$85,000

Notes:

1. The cash amount represents the exact amount of money on deposit in the bank at the end of the business day on December 31st.

2. Accounts Receivable is a new category, but it is typically present, and represents money that customers owe the business. It is an asset earned, but not yet received.

3. The value of supplies, as explained earlier, is the value you started with ($4000) and increased by the difference between supplies purchased and used ($5000).

4. The value of tools will remain unchanged if a constant inventory is maintained.

5. The value of the equipment has been reduced by the amount of depreciation assigned.
6. Real property will show appropriate depreciation of buildings; however, land is not depreciated in financial accounting.
7. The bank loan is being paid off and the unpaid principal at the end of the year is $45,000.
8. Accounts payable is another new category and represents money owed on trade accounts, for rent, or whatever on December 31st.
9. Contributed capital is unchanged.
10. Retained earnings were derived earlier.

The increase of $13,000 in owner's equity is reflected in the increase in the cash balance, the reduction in the bank loan, the difference between accounts receivable and payable, the increase in supplies, and the depreciation of the equipment.

Now, let's do some further analysis of this year's work. Was it a successful year or not? First of all, you were able to generate significant retained earnings, which is a healthy sign. But, what about rate-of-return? Before taxes, the net profit was $23,000. This equals the $41,000 net income before taxes less your $18,000 salary. The salary portion is compensation for services, not return on investment. After taxes, the net profit was $13,000. Both of these returns will be measured against a beginning owner's equity of only $25,000. Thus, your rates-of-return were

$$\frac{\$23,000}{\$25,000} = 92\% \text{ before taxes} \qquad \text{or} \qquad \frac{\$13,000}{\$25,000} = 52\% \text{ after taxes}$$

Not bad! Now, why didn't we calculate the return using the total value of assets for the denominator? The reason is that you have only $25,000 invested in those assets; others own the rest. You are paying the bank for the use of their money by paying them periodic interest and part of your gross income went for this purpose. In addition, your trade accounts (accounts payable) are providing indirect financing. Next year your investment will be $38,000 because of the retained earnings and that will become the denominator for next year's rate-of-return calculations. Hopefully, the business will continue to grow and owner's equity will increase each year. If a bad year comes along, owner's equity will decrease. You may find that these high returns on investment will never be repeated. Successful new businesses often show high growth rates in their early years but gradually slow down as the value of owner's equity increases. In this first year of business you have assumed considerable risk by going into debt and gambling on work by bidding on projects. These returns are fair in view of this risk. As your company becomes more stable and financially strong, the risks will be lessened and the rate-of-return should come down. The larger construction firms in the United States show an average *after-tax* return on investment around 15%. With federal corporate tax rates of 46% on taxable income over $50,000 (offset somewhat by investment tax credits) and with state taxes, where applicable, it appears that a before tax rate-of-return of about 30% is a reasonable target. Our discussion of rate-of-return is continued in the next section of this chapter.

Our return of $13,000 after taxes also can be compared to the total work performed of $200,000. In this case, this ratio (called *profit margin*) works out to be 6.5%. This figure is useful but not as significant as the return on investment calculation. Total business volume figures are misleading. If a $100,000 project is $90,000 labor and $10,000 materials it is quite different than one which takes only $5000 to install a $95,000 piece of machinery. Some businesses operate on a rate-of-return on business volume of only a few percent (discount stores, large food stores, etc.).

The key point of this section is that it is the owner's investment in the business from which measurements of fair rate-of-return are best calculated. This investment is found from the balance sheet of the firm and includes contributed capital and retained earnings (in the case of a corporation, contributed capital would be replaced by sale of stock).

RATE-OF-RETURN

In the discussion of balance sheets, rates-of-return were calculated for the first year of operation of the new contractor. In its basic form

$$\text{Rate-of-return} = \frac{\text{Net profit}}{\text{Investment}}$$

As illustrated in the previous section, the rate-of-return may be expressed before or after taxes; each method has its uses. The point is that when someone speaks of rate-of-return, the term must be defined. If you deposit money in savings account, you are told that you will earn interest at a rate of perhaps 6%. This return is taxable (before tax) income since the interest you earn on those deposits is subject to income tax. If your tax rate is 20%, you will pay $1.20 for each $6 earned so your rate-of-return on an after tax basis is only 4.8%. If you have the option of buying some taxable commercial bonds which yield 8% or some tax-exempt municipal bonds yielding 5.9%, which should you buy? The answer is: It depends on your tax bracket. If your tax bracket is 20% you will pay $1.60 on each $8 earned on the commercial bonds giving you a net return of 6.4% which is still better than the 5.9% you would earn on tax-exempt bonds. However, if you are in a 30% tax bracket, your return after taxes on the commercial bonds is only 5.6% and the municipal bonds are the better choice.

In the construction business, you will be making many financial decisions which require you to choose a course of action from among a number of possible courses. In most cases, you will evaluate these in terms of what your return will be and you must decide to use either a before tax or after tax position in your evaluation. In the example involving the bonds, it was the after tax return that was significant. When comparing returns on options which do not include any with tax-exempt features, the before tax approach is a bit simpler but after tax can be used.

The term *risk-free investment* refers to "sure-thing" types of investments. This is generally considered to mean only US Government securities. From this comes the term *risk-free rate-of-return* which, as you might expect, is the rate-of-return you could realize if you invested in US Government securities such as savings bonds, treasury bills, and other government bonds. The interest rate on savings bonds is approximately 6% with the rates on other government securities being higher and variable depending upon the money market at the time. These are before tax rates. Savings accounts in banks and savings and loan associations are also equivalent to risk-free investments since they are insured by the US government up to certain levels. Their rates range from about 5% upwards depending on the size of the account and its period of deposit.

The significance of risk-free rates is that they establish a datum from which one can evaluate other investment opportunities. Returning to the situation where you were starting a small contracting business, you could have decided to forget the whole thing and leave your $20,000 in a savings account at 6% interest. However, your net profit before taxes in that case would have been only $1200, not $23,000, and you would have had to find some other way to earn an $18,000 salary. If the contracting business had not been so successful and your earnings before taxes (after deducting your salary) had yielded a rate-of-return of only 4%, you would look at the situation and ask yourself why you wasted your time and strained your nerves working as a contractor when you could have earned more by leaving your savings in a savings account and getting another job. In other words, you definitely want a rate-of-return that is greater than a risk-free rate. In fact, because of the risks you are taking, you want a great deal more. This is a key point—your rate-of-return should be the sum of the risk-free rate-of-return plus an amount which properly compensates for the risk of doing business. We noted that the very large construction companies seem to operate with a before tax rate-of-return in the 30% range so this probably represents about the lowest you should strive for. A smaller contractor should seek a higher yield. This point will be put to use in Chap. 14.

TIME-VALUE OF MONEY

If someone asks you whether you would rather be given a dollar today or wait until a year from today and get that dollar, you would choose the option of receiving the dollar today. Why? Because, money has a time value. It can be put to work for you to earn interest or otherwise yield a return. Also, with inflation, dollars lose value with time.

Time-value-of-money considerations enter into any business transaction which involves time. If you were to buy two widgets today at $1 apiece, resell one of them today for $1 and sell the second one for $1 in 6 months, you have lost money. On the first widget sale you recovered the item value as you bought it. On the second widget, you lost 6 months interest on the $1 you invested in the widget. This time-value of money is another cost factor that you, as an estimator, must account for in determining true cost as opposed to apparent cost.

Time-value is accounted for mathematically by the equation

$$\text{Value at end} = (\text{Value at beginning}) (1 + i)^n$$

where: i = interest rate per period expressed as a decimal
n = number of periods

Thus, if you have $1000 today and wish to know what its value would be in 4 years if it is deposited in a 6% interest account, the calculation would be

$$\text{Value at end} = (\$1000) (1 + 0.06)^4 = \$1,262.47$$

Normally, interest rates are expressed as an annual rate, such as 6% per annum. When dealing in periods less than a year, the rate for the period in question is found by using either of these formulas

$$\text{Rate for period} = \frac{\text{Length of period in months}}{12 \text{ months}} (\text{Annual interest rate})$$

or

$$\text{Rate for period} = \frac{\text{Number of days in period}}{360 \text{ days}} (\text{Annual interest rate})$$

In the second formula, every month is equal to 30 days.

Now let us put some of these relationships into practice. Assume that a contractor goes to a local building supplier on the 1st of the month and purchases $5000 of material for cash for installation in a project. These materials are installed during the next 4 weeks. At the end of the month the contractor prepares a claim for progress payment from the client and includes the $5000 worth of materials on this claim. The architect monitoring the project validates the claim and sends it to the client for payment. The client pays the contractor on the 15th of the following month. Here, the contractor has had $5000 of company money tied up in supplies for 45 days. At the very least, it could have been deposited in a risk-free account for that period and its value would have been

$$\text{Effective rate for period } = \frac{1.5 \text{ months}}{12 \text{ months}} (6\%) = 0.75\%$$

$$\text{Value at end of period } = (\$5000)(1.0075) = \$5,037.50$$

The contractor effectively has lost $37.50 on this turnover. A better choice would have been to open a trade account (charge account) with the building supplier. In some cases these trade accounts offer no discount on the purchase but allow you 30 to 60 days to pay without interest. If the contractor had such an account, the materials originally could have been charged and paid for when the contractor was paid. There would be no loss. Other suppliers use what is called *trade credit*. Their terms will be something like "2/15 net 30" which means that you get a $2 discount per $100 purchased if you pay the bill within 15 days. There is no discount after the 15th day and the account becomes overdue after the 30th day and may be subject to carrying charges or collection action. If the contractor purchased the materials on the 1st and paid for them on the 15th the bill would be reduced by the discount making the actual payment $4900. If the bid included an allowance of $5000 for the materials, the contractor would recover this amount from the client on the 15th of the following month and will have more than recovered costs plus the time-value of those costs. This time the contractor has had $4900 tied up for only 1 month so the arithmetic would show

$$\text{Effective rate for carrying period } = \frac{1 \text{ month}}{12 \text{ months}} (6\%) = 0.50\%$$

$$\text{Value at end of period } = (\$4900)(1.005) = \$4924.50$$

In addition, the contractor had the use of the $4900 from the first to the fifteenth of the month when the goods were paid for. This was worth

$$(\$4900)\left(\frac{\frac{1}{2}\ \text{month}}{12\ \text{months}}\right)(0.06)\ =\ \$12.25$$

In this case the contractor has taken advantage of time-value by holding off payment for the materials until the last day of the discount period so as to reduce the carrying period while still getting full discount. Then, the contractor added to the profit by originally allowing $5000 in the bid price for these materials and this amount can properly be claimed in the current progress payment. It is the combination of actions such as these in the total business process that often makes the difference between contracting success or failure.

CASH FLOW

It is quite unlikely that a construction contract would be written to allow a client to wait until project completion and acceptance to pay for it. The more normal way of project financing is through progress payments. This means that the contractor is paid an amount each month, or other agreed upon period, for work performed during that period. In some cases a percentage is retained by the client as leverage over the contractor until the project is finally accepted. Similarly, a bank or other lending institution that is financing a project will not give a builder a lump-sum loan at the beginning of a project since there is no collateral at that point. Instead, they will release funds periodically based upon work accomplished. This periodic approach results in a funding pattern called *cash flow*.

One of the purposes of planning and scheduling a project is to translate the work schedule into a cost schedule or budget. To explain by example, assume that a small building is to be erected and the schedule for construction looks like that given in Fig. 3-1. The contract price for this project is $25,000 and the estimator has broken this down by work items. This information also is shown on the schedule. Assuming that the work is done on schedule, the contractor would be entitled to progress payments as shown below the figure.

For the client, this cash flow schedule would be used for budgeting to insure that money in adequate amounts would be available to meet payments. For the contractor, this schedule would be used for recruitment and payment of required labor, etc.

INFLATION

Inflation is another effect which time can have on money. When discussing the time-value of money earlier in the chapter, it was presumed

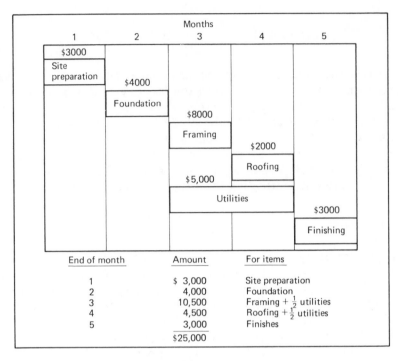

End of month	Amount	For items
1	$ 3,000	Site preparation
2	4,000	Foundation
3	10,500	Framing + $\frac{1}{2}$ utilities
4	4,500	Roofing + $\frac{1}{2}$ utilities
5	3,000	Finishes
	$25,000	

Fig. 3-1   Cash flow schedule.

that the added value generated by time was the result of the money working for the investor and that the investor would have more buying power at the end of the investment period than at the beginning. Inflation counters money time-value; in periods of severe inflation, it can completely overcome it. An investor in a savings account drawing 6% interest during a period of 10% annual inflation of the cost of living is effectively losing money. During the decade 1969–1978 the cost of construction rose at an equivalent compounded annual rate of 9%. In 1972 a leading brand of dozer cost $68,785; the same model cost $127,425 in 1977. Thus, in a 5-year period the price of that type of equipment rose at an equivalent compounded annual rate of more than 13%. Obviously, a contractor bidding on a project of any duration must think about the implications of inflation. It can be ignored only if suppliers of all construction resources, including labor, can give guaranteed prices. During 1979 the price of fuel more than doubled in many areas. No fuel supplier can be expected to commit to a set price for a significant period. The same will apply to almost any vendor. Thus, the contractor is left with two options when bidding on a project. First, the contractor may refuse to sign any contract that does not contain protective escalation clauses (see Chap. 4). Second, inflation factors must be included in all cost estimates.

Inflation Considerations in Estimating

In preparing an estimate, the estimator first costs the project as if it were to be built entirely today. The estimator then projects current costs of each of the activities to the time that they occur on the schedule using an inflationary rate based on his or her best judgment. The cost of this activity then becomes the inflated cost and will be used in the estimate. This is repeated for all activities or groups of activities at a particular time. The sum of all these costs becomes the bid. This procedure has the effect of adding different value dollars but, in contracting, it is actual dollar flow over the counter that is needed for the bid—these dollars are not reduced to any particular year. To show how this procedure works in practice, assume that there is a project scheduled as shown in Fig. 3-2 with costs of activities calculated in dollars at the time of the bid.

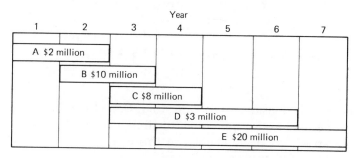

Fig. 3-2 Cash flow at current prices.

The second effort is to estimate how the money will be expended within each activity. For the example of Fig. 3.2 assume that the expenditures for each activity for each year (at current prices) are planned as follows:

	$ Millions						
Activity	Year 1	Year 2	Year 3	Year 4	Year 5	Year 6	Year 7
A	0.5	1.5					
B		4.0	6.0				
C			3.0	5.0			
D			0.5	1.0	1.0	0.5	
E				3.0	7.0	7.0	3.0
Totals	0.5	5.5	9.5	9.0	8.0	7.5	3.0

These yearly totals are then inflated using an assumed inflation rate. In this case assume 9%. To do this we use the time-value formula discussed earlier in this chapter and project to the midpoint of each year. We can do this by dealing in 6-month periods and use whole number exponents, or we can deal in 1-year periods and use fractional exponents. For example, the midpoint of year 3 can be handled with either of the following expressions

$$\text{Present value} \times (1.045)^5 \qquad \text{or} \qquad \text{Present value} \times (1.09)^{5/2}$$

In the first case we are using five 6-month periods (at 4.5% inflation per 6-month period); in the second, we are using $2\frac{1}{2}$ years at 9% per year. To continue with the example, assume the second method is used.

Year 1: ($0.5 million) $(1.09)^{1/2}$ =   0.52 million
Year 2: ($5.5 million) $(1.09)^{3/2}$ =   6.26
Year 3: ($9.5 million) $(1.09)^{5/2}$ = 11.78
Year 4: ($9.0 million) $(1.09)^{7/2}$ = 12.17
Year 5: ($8.0 million) $(1.09)^{9/2}$ = 11.79
Year 6: ($7.5 million) $(1.09)^{11/2}$ = 12.05
Year 7: ($3.0 million) $(1.09)^{13/2}$ =   5.25

Bid price = $59.82 million

This bid price is considerably higher than the $43 million calculated at current prices and again emphasizes the importance of evaluating potential inflation in bid development. In this example, we dealt with total work packages and used a single rate of inflation. In practice, labor, material, and equipment costs will inflate at different rates so an estimator may choose to divide work packages into resource categories and inflate each at a different rate.

Now, imagine that the project just described fell behind schedule a year or two. The deferment of any activity will cause an inflationary increase in its cost.

DEPRECIATION

Depreciation expense recognizes that a piece of equipment or a structure loses its value with time due to a combination of wear and obsolescence. This loss of value is a business expense which must be accounted for in the financial records of the firm and included in direct or indirect charges allocated to projects.

Equipment Depreciation

The income tax code permits an owner of equipment to calculate its depreciation in one of three ways for tax purposes: straight line, sum-of-the-digits, or declining balance. The code also specifies a range of years over which the depreciation can be carried. In the case of most construction equipment, this range is 4 to 6 years. The code does allow exceptions if an item is to be used under operating conditions so severe as to shorten its life.

Straight-Line Method. This method assumes that a piece of equipment loses value at a uniform rate over its economic life. For example, a piece of equipment originally costing $20,000, with an economic life of 5 years and an estimated salvage (or trade-in) value of $5000 at the end of the 5 years would have its annual depreciation calculated as

$$\text{Annual depreciation} = \frac{\text{Purchase price} - \text{Salvage value}}{\text{Economic life}}$$

$$= \frac{\$20,000 - \$5000}{5 \text{ years}}$$

$$= \$3000 \text{ per year}$$

Sum-of-the-Digits Method. This is an accelerated depreciation method in that it permits an owner to take high depreciation in the first year and successively less each of the remaining years. To use this method the user must first calculate the sum of the digits representing the years included in the economic life. For example, for a 5-year life, the sum is $5 + 4 + 3 + 2 + 1 = 15$. This sum can be determined more quickly by use of the formula

$$\text{Sum of digits} = \frac{n(n+1)}{2}$$

where n = economic life, in years.

The depreciation for the first year is found by applying a fraction that has as its numerator the digit representing the economic life and as a denominator the sum of the digits. For each following year the numerator is reduced by 1 as follows:

$$\text{First year:} \quad \frac{n}{\text{Sum}} \times (\text{Purchase price} - \text{salvage value})$$

Second year: $\dfrac{(n - 1)}{\text{Sum}}$ X (Purchase price – salvage value)

Third year: $\dfrac{(n - 2)}{\text{Sum}}$ X (Purchase price – salvage value)

Using the data from the previous example, depreciation would be

First year: $\dfrac{5}{15}$ X ($20,000 – $5000) = $5,000

Second year: $\dfrac{4}{15}$ X ($20,000 – $5000) = 4,000

Third year: $\dfrac{3}{15}$ X ($20,000 – $5000) = 3,000

Fourth year: $\dfrac{2}{15}$ X ($20,000 – $5000) = 2,000

Fifth year: $\dfrac{1}{15}$ X ($20,000 – $5000) = $\underline{1,000}$
$$\$15,000$$

Note that the total depreciation is still only $15,000, the difference between original purchase price and estimated salvage value but the amount taken in individual years decreases with time.

Declining-Balance Method. This is a second form of accelerated depreciation available for tax purposes. The formulas applicable are as follows. For equipment purchased new

$$\text{Depreciation} = (2.0) \times \frac{(\text{Remaining book value})}{(\text{Economic life})}$$

For equipment purchased used

$$\text{Depreciation} = (1.5) \times \frac{(\text{Remaining book value})}{(\text{Economic life})}$$

In these formulas, remaining book value is equal to original purchase price less accumulated depreciation to date. Note that salvage value is not included in either formula. In using this approach, an item is depreciated until estimated salvage value is reached, even if that point is reached before the expiration of the economic life. Using the previous data, depreciation under the declining balance approach (assuming new equipment) is

$$\text{First year: Depreciation} = (2.0) \times \frac{(\$20,000)}{5} = \$8000$$

Remaining book value = \$20,000 - \$8000 = \$12,000

$$\text{Second year: Depreciation} = (2.0) \times \frac{(\$12,000)}{5} = \$4800$$

Remaining book value = \$12,000 - \$4800 = \$7200

$$\text{Third year: Depreciation} = (2.0) \times \frac{(\$7200)}{5} = \$2880$$

Remaining book value = \$7200 - \$2880 = \$4320*

*Since \$4320 is less than salvage value of \$5000, change the depreciation to that amount which brings remaining book value to \$5000.

$$\text{Depreciation} = \$7200 - \text{Salvage value} = \$2200$$

Remaining book value = \$7200 - \$2200 = \$5000

Fourth and fifth years: No depreciation allowed since full \$15,000 already exhausted.

Since a business can choose any one of the three depreciation methods for income tax purposes, a contractor must evaluate alternatives considering the time-value of money, the need for business deductions in certain years, and the expected tax bracket of the business each year. However, once a method is chosen, it normally must be continued for the life of that piece of equipment.

Depreciation accounting for income tax purposes is one of three accounting situations when depreciation must be calculated by a contractor. The second is for bookkeeping and financial reporting. The third is for determining charges for equipment use on projects.

The depreciation method used for bookkeeping and financial reporting does not have to be the same as that used for income tax purposes nor does it have to be one of the three methods discussed for income tax. In fact, the frequent incompatability of accepted accounting principles with income tax law often dictates two sets of books. One rule does apply to both, however—an item cannot be depreciated beyond its purchase price or its purchase price less salvage value.

Whatever the method of accounting, the purchase price to be used in the two depreciation calculations is the delivered price including taxes, dealer preparation charges, and transportation costs. It does not include license fees. Also, in the case of rubber-tired equipment and trucks, a common procedure is to deduct the cost of tires and handle tires as a separate operating cost.

The handling of the depreciation component of cost in the development of charge rates for use of construction equipment is independent of the method or methods used for income tax and financial records. This subject will be discussed in detail in Chap. 10.

Building Depreciation

Buildings, but not the land on which they are located, can be depreciated and the depreciation taken as a business expense. Structures are normally depreciated over a period of 10 to 40 years and the straight-line procedure is employed.

INVESTMENT TAX CREDIT

The *investment tax credit* (ITC) is a form of incentive for business investment in capital equipment such as cranes, bulldozers, and welding machines that a contractor might purchase. It is a one-time benefit taken in the year of equipment purchase. In most cases the benefit can be taken only by the buyer of the equipment; an exception can occur in the leasing of equipment in that the lessor may pass on the ITC to the lessee (see Chap. 10). The calculation of ITC is as follows:

$$\text{ITC} = (\text{Purchase price}) \times (\text{Multiplier})^* \times (0.10)^\dagger$$

Most construction equipment is allowed an Asset Depreciation Range of 4 to 6 years by the IRS. The ITC is used to offset taxes due dollar for dollar. Thus, it is more valuable than a simple business deduction. For purposes of explaining both the calculation of ITC and its handling in income tax calculations, assume that a corporation has gross income of $2,000,000, business expenses (deductions) of $1,500,000 and purchased $450,000 of equipment with a 5-year life during 1979. In 1979, corporate tax rates were 17% of the first

*Multiplier is zero if life span is 2 years or less
Multiplier is $\frac{1}{3}$ if life span is 3 or 4 years
Multiplier is $\frac{2}{3}$ if life span is 5 or 6 years
Multiplier is 1 if life span is 7 or more years
†Based on current 10% ITC rate. Rate has been 7% in some years and nonexistent in others.

$25,000, 20% of the next $25,000 and 46% of all remaining taxable income.

$$\text{ITC} = \$450,000 \times \tfrac{2}{3} \times 0.10 = \$30,000$$

Gross income	$2,000,000
Less: Tax deductions	1,500,000
Taxable income	500,000

Taxes owed:

(0.17) ($25,000)	=	$4,250
(0.20) ($25,000)	=	5,000
(0.46) ($450,000)	=	207,000
		$216,250

Less ITC	30,000	
Taxes to pay	$186,250	186,250
Earnings after taxes		$313,750

SOURCES OF CONTRACTOR FUNDING

Sources of contractor funding can be divided into two general categories: direct and indirect. Direct contractor funding includes owner-contributed capital, retained earnings, and loaned capital. Indirect funding includes trade accounts, installment purchases, and equipment leasing and renting.

Contributed Capital

The owner or owners of a construction contracting organization will have contributed some combination of money, equipment, tools, supplies, and real property to the business when the firm initially goes into business. In many cases, this is the only funding source during early months or years of operation since the company has yet to establish a financial track record and may be considered too big a risk for institutional lending, especially during periods of tight money.

If the new construction firm is being organized as a corporation, all contributed capital (whether money or property) is exchanged for shares of stock in the corporation. If the company is closely held (owned by the principals in the company or family members) the amount of available capital is limited. If the company goes public and offers shares for sale to anyone, their limit on funding is entirely a function of their ability to attract money.

In the case of a single proprietorship or partnership, the owner (owners) may contribute additional capital to the business at any time as a means of increasing their equity in the firm. In the case of a corporation, additional equity capital is obtained through sale of additional shares of stock. When a single proprietorship or partnership is converted to a corporation, the owner's equity is converted into an appropriate number of shares of stock.

Retained Earnings

As discussed in the section on balance sheets, retained earnings are derived from the profits of the business after all expenses and taxes have been paid. This net profit after taxes will be distributed to the owners as dividends or will be reinvested in the business as retained earnings. These retained earnings plus the owners' contributed capital constitute their equity in the firm. Through continued addition of retained earnings, the owners' equity will grow.

Borrowing

Banks.   Banks are the most common source of loaned capital for smaller construction firms and all construction firms probably utilize bank loans at some point in their business. Banks are primarily in the short-term loan business; long-term loans are the province of savings and loan associations, insurance companies, pension funds, etc. Contractors as a class are considered high risk. Only the largest firms with excellent financial history obtain preferred customer status. Thus, a contractor looking for bank support can expect detailed examination of the firm's past record of performance, the quality of its management group, the type of work engaged in, plus its current financial situation. Bank financing can take these forms

1. *Line-of-Credit.*   A line-of-credit (LOC) is a borrowing limit that a bank agrees to after a financial analysis of a firm. In effect, it is an open charge account for a specific period so that a contractor need not go back to the bank each time the firm needs money for an expenditure. However, a line-of-credit is opened only if a contractor has something definite in mind such as funding for a project in hand or one on which they want to bid. There is one condition associated with lines-of-credit and that is a requirement to maintain a *compensating balance* on deposit in the bank during the period that the line-of-credit is effective. A compensating balance is expressed as a percentage of the line-of-credit outstanding, usually in the range of 15 to 25%. For example, a bank may specify a 20% compensating balance on a line-of-credit of $100,000. This means that the borrower must maintain a checking account balance of up to $20,000 (20% of amount on loan) in that bank at all times that the line-of-credit is effective. It also means that

the borrowers will not have the full line-of-credit available for use if they are unable to maintain the required compensating balance from other funds and part of the line-of-credit must be used to maintain this balance. Of course, there is a basic interest charge on all money currently on loan, including any in the checking account that was drawn from the LOC. Since every contractor must maintain an amount of funds in a checking account to meet everyday expenses, the compensating balance may not prove to be a hardship. The line-of-credit is designed to cover short-term debts of the constructor and would not be used for major equipment or real estate purchases. The bank also normally requires that the line-of-credit be paid off (reduce loan balance to zero) every 6 months or so.

A construction company may have lines-of-credit with several banks, especially if its projects are located in different communities. It is always good public relations to do banking business in the community of the project.

2. *Collateral Loan.* Some banks will give loans for purchase of equipment in much the same way as they give auto loans. They may require that the borrower pay a certain percentage of the purchase price. Interest rates usually increase with the percentage of funding provided by the bank. In all cases, the bank holds a lien on the purchased property until the loan is paid off. Some banks refuse to give equipment loans since the buying and selling of construction equipment is not as structured as is the case with automobiles and the risk is considered to be too great.

3. *Individual Project Loan.* This type of loan is designed to provide funds as needed by a contractor during the construction of a speculation project. In operation, the bank commits funds to the contractor to purchase the necessary materials and to meet the payroll for a project as it progresses. The complete loan comes due shortly after project completion. Should the contractors intend to keep the properties for rental or be unable to sell them as planned they would be expected to obtain mortgage financing from a savings and loan or other long-term lender to pay off the bank.

Mortgage Lenders. In this category are savings and loan institutions, insurance companies, pension funds, and other groups who regularly invest in real estate through mortgage loans. As a general rule, these organizations take over financing of completed projects, not their construction. Terms of loans are in excess of 10 years and more often in the 20 to 30 year bracket. The lenders protect themselves by means of a mortgage on the property covered by the loan.

Long-Term Loan. Construction firms with extremely strong financial positions and with recognized quality management may attract investment funds in the form of loans from insurance companies, pension funds, investment bankers, etc. These loans generally contain very strict conditions on uses of retained earnings, limitations on dividends, etc., as a means of protecting the investor.

Bonds. Bonds are a special form of long-term borrowing. Because of the regulations and costs associated with marketing a bond issue, this option is available only to the largest organizations. Bonds are sold almost invariably with a face (par) value of $1000 and with a specific interest (coupon) rate. The buyer of the bond may pay more or less than the face value of the bond but is assured of receiving exact face value of the bond on redemption and of receiving interest annually at the stated rate computed against the face value. Bond maturity periods are 10 years or longer. The buyer of the bond is strictly a lender of money; the buyer acquires no equity interest in the firm unless the bond carries some form of conversion privilege which allows exchange for shares of stock.

Trade Account

A trade account is a charge account. It may or may not involve discounts for early payment. Some trade accounts will carry interest charges from day of delivery; for others there is no interest if payment is made within a specified period. In the earlier discussion of the time-value of money a form of trade account was mentioned. A trade account is actually a form of short-term loan because the contractor is being permitted to use goods and services in advance of payment therefor.

Installment Purchase

This form of purchase is the same as that used for merchandise and automobile financing by typical consumers. An item is purchased, with or without down payment, and paid off by means of periodic installment payments which include principal and interest. As with a trade account, the user of an installment purchase is borrowing money in the form of goods or services and paying for them at a later date.

Equipment Leasing and Renting

Many of the equipment needs of a contractor can best be met by leasing or renting the items. This subject is discussed in considerable detail in Chap. 10. These options are an indirect form of financing for a contractor. In some lease contracts, provision is made for eventual purchase by or transfer of title to the contractor leasing the item. Such lease contracts are technically equivalent to installment purchase.

SUMMARY

This chapter has roamed over a wide range of subjects in the business and financial management areas. The coverage of each was brief but the objective has been realized if you are now on speaking terms with them. In following chapters these subjects will be directly related to various cost-estimating actions.

The contractor who enters the construction business as a great engineer and builder but without knowledge of business and finance is the contractor with a high risk of failure. This chapter will not make you an expert in these areas, but it can alert you to the complexity of the business world and should cause you to pursue further inquiry into those areas where you are weak should you become a contractor or a key member of a contractor's staff.

REVIEW

Exercises

1. John Smith has entered the contracting business as a single proprietor. He contributes $35,000 in cash and property to start the firm. He also obtains a loan from his sister-in-law for $10,000. During his first year of operation he has gross earnings of $200,000, expenses (other than his salary) of $170,000 and he pays himself a salary of $20,000. Assume his tax rate is 35% after personal deductions of $7000. He leaves all net profit in the firm as retained earnings. What is his equity at the end of the first year of operation?

2. In Exercise 1, John Smith has equipment purchased at the beginning of the year for $10,000 and which is depreciated at a rate of $1800 per year. He also owns 1 acre of land that was part of his contributed capital and which is valued at $5000. During his first year of operation he repaid $2000 of the $10,000 loan. He has a supply yard inventory of $7000 and has no outstanding trade accounts. Prepare his balance sheet at the end of the first year.

3. In Exercise 1, what is
   a. John Smith's before tax return on investment?
   b. John Smith's after tax return on investment?

4. If $20,000 is deposited in an account at 7% annual interest, what is the value of that account after 5 years if interest is left on deposit?

5. A contractor deals with a supplier whose terms are "2/10 net 30." The contractor purchases $1000 worth of supplies on the 1st day of the month.
   a. What will the contractor pay for these supplies if he or she pays cash on delivery?
   b. What will the contractor pay if the bill is paid on the 10th of the month?
   c. What will the contractor pay if the bill is paid on the 15th of the month?

6. A long-term project is composed of these major work packages
   Package A: Valued at $20 million to be accomplished during years 1 to 4.

Package B: Valued at $10 million to be accomplished during years 1 and 2.

Package C: Valued at $16 million to be accomplished during years 3 to 6.

Package D: Valued at $8 million to be accomplished during years 4 to 7.

Assume that work performed within each package is spread evenly over the years indicated; also, that values given are current prices. The anticipated average annual rate of inflation is 8%. Develop yearly cash flow figures and translate these into a bid figure for the project.

7. A piece of equipment cost $20,000 5 years ago. The current cost of a comparable item is $35,250. What has been the average annual rate of price inflation for that equipment item?

8. A piece of equipment has an economic life of 5 years. Delivered cost is $55,000 and estimated trade-in value after 5 years is $15,000. Calculate the depreciation schedule using
   a. Straight-line
   b. Sum-of-digits
   c. Declining-balance (new equipment)

9. What is the investment tax credit for the piece of equipment in Exercise 8? Use a 10% ITC rate.

10. A contractor obtains a line-of-credit from a local bank in the amount of $75,000 with a requirement for a 20% compensating balance. The contractor is able to maintain a $10,000 checking account balance from normal operations. If the interest rate on the line-of-credit is 12% and the contractor uses the line-of-credit for purchase of $60,000 in supplies and does not repay this amount for 3 months, what interest is owed?

Questions in Review

1. Why would a bank or other lending institution hesitate to provide 100% of the funds needed to start a business?

2. How is rate-of-return measured?

3. What is the basic equation for a balance sheet?

4. What are the sources for a firm's assets?

5. What is meant by the statement that depreciation is not a cash flow item?

6. Is there any difference between retained earnings and net profit after taxes?

7. What two considerations determine a fair rate-of-return in a business?

8. Why is development of a cash flow pattern important in the estimating of a project?

9. What capital sources for a business are reflected in the owner's equity portion of a balance sheet?

10. What type of borrowing may be available from banks?

11. If you were to build a house for yourself, describe the probable method of financing it during construction and after occupancy.

12. Describe some passive methods of contractor financing.

Discussion Questions

1. If you were a banker considering a loan to a contractor, what would you look for to help you decide whether the contractor is a good risk, if
   a. He or she were just starting out?
   b. He or she had been in business several years?
2. Many contracts call for retention of about 10% of all progress payments by the client until satisfactory completion and acceptance of the entire project. If you were bidding on such a contract, how would you allow for this provision?
3. Equity financing refers to obtaining capital through contributions from the owners or sale of stock in the case of a corporation. Debit financing refers to sale of bonds or other borrowing. What are the advantages and disadvantages of each method?

CONTRACTS AND CONTRACT DOCUMENTS

4

INTRODUCTION

The entry of two parties into a contract for provision of goods and services is a serious matter whether the contract be oral or written. In construction, the contract is particularly important since the definition of the project and the obligations of a contractor in building it are difficult to define in absolute fashion. Thus, over the years, the procedure for construction contracting has become more formalized and structured as successively improved contract forms and procedures have been tried and tested in courts of law. This chapter is devoted to an overview of contract types and contract documentation as further background for the cost-estimating process.

TYPES OF CONSTRUCTION CONTRACTS

Lump-Sum

The lump-sum (*also called "fixed-price" or "hard-dollar"*) contract specifies that the contractor will construct the facility described by the plans and specifications for a fixed amount of money. The amount of money is that determined by the contractor after reviewing the contract documents and includes the contractor's estimate of all direct costs, indirect costs, overhead, and profit. The contractor is not obligated to reveal to the client the calculations used to arrive at the quoted price and the client has the option of selecting or rejecting a bid. In Europe, a common third option is to negotiate after bids are in to further lower price but this is not common practice in the United States. The contractor assumes all risk on a lump-sum contract. The contractor is gambling that he or she can construct the structure at a cost that does not exceed the bid cost. If costs do go over, the contractor is eating into his or her profit margin or may actually lose money. Obviously, the ability of a contractor to make an accurate estimate is of paramount importance.

On a project involving public funds, governmental procurement regulations normally require award to the lowest responsive bidder following an open advertisement for bids. Under this system, the invitation to bid is made public through advertisements in newspapers and trade magazines, and through other publicity devices. Interested contractors request copies of the contract documents and prepare their bids which must be submitted by a certain hour of a certain day at which time bids are opened in the presence of all contractors who wish to attend. Each bid is read and an announcement is made of the apparent low bidder. This bid is subsequently examined to insure that there is no bidding error, that all requirements of the invitation have been met, and that the contractor has the apparent financial, technical, and physical capability to execute the project. The client may choose to reject all bids if the bid figures exceed an amount that the client considers reasonable or affordable. Bidders are not reimbursed for any expenses incurred in preparing their bids.

Private clients can be selective and may choose to negotiate a contract price with a single contractor although they are more likely to select several contractors and invite bids.

The amount paid the contractor by the client will vary from the bid amount only as a result of changes to the project from that described in the contract documents. This is a touchy legal subject so the contract documents will describe in considerable detail through a *changes clause* what can be changed, how it is changed, and how price adjustments will be handled.

The plans and specifications for a lump-sum bid project must be essentially complete to enable all bidding contractors to analyze the project in detail before preparing their estimates.

Unit-Price

This type of contract is a special form of lump-sum contract. A lump-sum contract quotes a package price for all included components; a unit-price contract breaks the project into units and shows a bid price on each unit. A unit, as used here, has a broad definition. It can mean a complete structure, such as a bridge, that is part of the larger project or it may be a unit of work measurement, such as linear foot of ditch. An example can best explain the approach. Assume that a section of road that is to be built includes subgrade haul and fill, base course haul and placement, and an asphaltic concrete surface. In addition, two box culverts and one bridge are to be part of the project. The total project will be described on the bid form by a listing of work items selected by the client. Some of these work items may be self-explanatory while others are not since they encompass several work packages that can be found in the plans and specifications but are not listed at all on the bid form. For our example, assume the client (or the client's engineer) describes the project on the bid form by these six items.

Item	Unit*	Quantity
Mobilization	LS	1
Excavation, right-of-way	CY	28,000
Excavation, borrow	CY	12,000
Asphaltic concrete, in place	SY	15,000
Box culverts	EA	2
Bridge	LS	1

*LS, lump sum; CY, cubic yard; SY, square yard; and EA, each.

Looking at the list you might wonder where such items as spreading and compacting the fill, or haul and placement of the base course are. True, they are not shown, yet the bidder must consider these and every other possible work task required to complete the project but must account for their cost in the six items given on the bid form. This applies also to overhead and profit. To do this, the contractor will develop estimating work sheets in about the same way as he or she would do for a lump-sum contract. Once everything is identified and costed, the contractor will transfer the cost of each work package or cost item to one of the categories shown on the bid form. He or she may transfer the whole cost to a single item or may distribute it over some or all the items. For example, all costs of cutting, hauling, spreading, and compacting common excavation along the right-of-way would probably be assigned to the category "Excavation, right-of-way." The same would be done for "Excavation, borrow." Under the item "Asphaltic concrete, in-place" would go all costs of procuring, hauling, placing and compacting the base course plus all costs associated with placing the pavement itself. Box culverts would be individually estimated and their costs averaged. The bridge would be estimated as a lump-sum structure. Overhead, indirect costs, and profit would then be spread over the entire package. A fee also would be listed for "Mobilization" which reflects both mobilization and demobilization costs. Once an amount has been determined for each item on the bid form, it is then divided by the quantity shown on that bid form to give the unit price. In our example, assume that total data works out to be

Item	Unit	Quantity	Unit Price	Extended
Mobilization	LS	1	18,000.00	18,000.00
Excavation, right-of-way	CY	28,000	5.00	140,000.00
Excavation, borrow	CY	12,000	6.00	72,000.00
Asphaltic concrete, in-place	SY	15,000	7.00	105,000.00
Box culverts	EA	2	10,000.00	20,000.00
Bridge	LS	1	180,000.00	180,000.00
		Extended total		$535,000.00

Now you can imagine, in a competitive bidding situation, that each contractor will come up with different unit prices, and they do. One contractor may be low on one item, but high on another. Thus, for purposes of awarding the contract, the unit prices are not significant—it is the extended price total that is used to determine the low bidder. (In Chap. 18, unbalanced bidding will be presented as a method by which contractors can adjust unit prices and extended totals for the purpose of giving them both a bidding advantage and an ultimate profit advantage.)

The unit prices become important in the payment process while the extended total is no longer used after the contract is awarded. The contractor will be paid the bid unit price for each unit completed. For example, if the number of cubic yards of borrow, as actually measured, turned out to be 13,000 instead of 12,000, the contractor would be paid $78,000 for that item of work. In the end, the total amount paid can be less than, equal to, or more than the extended total on which the bid was accepted.

With a unit price contract the plans and specifications must be essentially complete on any structures to be bid as lump sum (LS) within the bid. Other plans need only be in such detail as to enable the bidder to determine the unit cost of production. For example, such items as quantities of earthwork cannot be exactly defined and that is one reason for the existence of the unit price type of contract. Actual pay quantities will be measured as the work progresses.

Cost-Plus-Fixed-Fee

There are many occasions when a contractor is asked to bid on a project in advance of completion of its design as a means of saving time and corresponding cost or asked to bid on a project containing technology or design features new to the general construction industry in the area of the project. Contractors are not generally willing to provide a lump-sum bid on such work or, if they do, will heavily inflate the bid to reflect the risk entailed. This has led to the cost-plus-fixed-fee contract. In signing this contract the client agrees to pay all valid costs incurred by the contractor in its execution plus pay a fixed fee representing contractor overhead and profit. On a large project, such as a nuclear power plant, or on turnkey contracts, a cost-plus-fixed-fee contract is quite common between the client and the contractor. However, many of the subcontracts awarded by the contractor are advertised after plans on those portions of the project are well defined and the contracts will be lump-sum or unit-price so the client can still be gaining some advantages of lump-sum contracting. Cost-plus contracts shift most risk to the client.

The obvious disadvantage of a cost-plus-fixed-fee contract is the uncertainty of final cost. Many a project has been scrapped, slowed down, or scaled down to keep within budget or cash flow capabilities of the client. Another problem is definition of reimbursable costs; the contract must be carefully drafted to prevent a rash of conflicts between contractor and client.

Cost-Plus-Percent-of-Cost

This is a variation of the cost-plus-fixed-fee contract in that the fee paid the contractor is a percent of the cost of construction. This form is seldom used since it provides no incentive for the contractor to hold down costs or complete the job.

Cost-Plus-Fixed-Fee-with-Target-Cost

As another variation of cost-plus-fixed-fee, this contract adds a target cost for the construction. If the construction exceeds that cost, the contractor absorbs a portion thereof. If it is less, the contractor shares the savings with the client according to some predetermined formula. The target cost is determined through contractor–client negotiation. A good target cost is one which both parties agree has an equal chance of being overrun or underrun.

Escalation Clauses

During periods when labor and/or material prices are escalating at a rate which makes cost estimating particularly uncertain, escalation clauses can be inserted in any form of fixed price, unit price, or target price contract. Under the escalation concept, a fixed or target price is established using base prices for those items expected to escalate. Then, procedures are included for adjusting the fixed price or target price as these items escalate in cost.

CONTRACT DOCUMENTS

A client desirous of having construction work performed will publicly or privately seek contractors to perform this work by inviting them to bid on the project or to submit a proposal for doing the work. Upon selection of a contractor by the client, a contract is signed by the two parties and work can proceed according to its terms. This section will detail the various documents that can be part of this process.

Bidding Documents

Bidding documents include all instructions and forms that are part of the bidding process. Bidding documents will be accompanied by General Conditions, Supplementary Conditions, drawings, and specifications to enable contractors to determine the full scope of their obligations should they undertake the construction. Depending on the particular contract, bidding documents will include some combination of the following:

1. *Invitation for Bid (IFB).* This document officially requests the submission of bids. It gives a general definition of the project and its location, identifies where contract documents may be obtained, establishes the deadline for receipt of bids, and provides other general instructions. The IFB may be in the form of an advertisement in a trade magazine or paper or a letter addressed to contractors selected by the client.

2. *Instructions to Bidders.* These may be incorporated within the Invitation to Bid or presented in a separate document. They contain more detailed instructions concerning bid requirements. Typical included items are

    a. *Requirement for Bid Bond.* A bid bond is a form of third-party insurance. The surety providing the bond guarantees that the contractor being bonded will accept the contract if it is offered. A typical bid bond would be for an amount equal to 5 to 20% of the bid price. Should the contractor fail to accept the contract after bidding, the client would be reimbursed for the difference between that contractor's bid and the next higher responsive bid. Some clients will require or allow substitution of a certified check for the bid bond.

    b. *Requirement for Performance Bond.* The performance bond is a guarantee that the contractor will actually complete the contract. It is normally for 100% of the contract price.

Should the contractor default, the surety must take action to complete the contract. The bond is not purchased prior to contract award but a guarantee from a surety may be required at time of bid which states that they will issue the bond if the contract is awarded.

c. *Requirement for Payment Bond.* The payment bond guarantees payment of labor and suppliers of the contractor. It will be in range of 40 to 50% of the contract price. Failure of the contractor to meet these obligations requires surety assumption of the debts. Like the performance bond, the payment bond is not required at time of bid, but assurance of its availability may be required.

d. *Identification of Contract Forms to Be Used*

e. *Method of Obtaining Drawings and Specifications*

f. *Procedure for Administering Modifications Prior to Bid Opening Date*

g. *Announcement of Prebidding Conference*

h. *Procedure for Selecting Contractor*

i. *Special Requirements.* Included would be such items as handling of sales taxes on public projects; assurance by a union contractor of no adverse union action on an open-shop project; description of contractor's equal employment opportunity, quality assurance, or other programs; identification of intended subcontractors; description of technical qualifications and experience of firm; etc.

3. *Bid Form.* A bid form is a standard form used by a client and adapted to a given project. When filled in, it will be addressed to the client and signed by the contractor and state, in effect, that the contractor agrees to construct the project described by the plans and specifications, to comply with the general conditions, special conditions, and all other requirements contained in the bidding documents and to do the work for a given price (or schedule of unit prices, cost-plus-fee, or whatever). A blank form is included within the bidding documents.

4. *Contract Form.* The client will either attach a copy of the contract form to be used for the project or will identify a standard form that will be used.

5. *Schedule of Plant.* The client may request a list of equipment available to the contractor which will be used on the project. This is used in evaluating the contractor's capability to perform the work. A form will be provided for this schedule.

For a contractor's bid to be considered responsive, all bidding requirements established in the bidding documents must be met by the bidder. If not, the client may reject the bid.

General Conditions or General Provisions

This document contains the generally accepted practices for administration of a contract for the type of facility involved. Trade or professional associations such as the Associated General Contractors of America (AGC) and the American Institute of Architects (AIA) have developed standard sets of general conditions that are available for use by clients. These standardized documents have been developed over many years and have been tested through use and court decisions so are definitely recommended in preference to General Conditions locally developed. The AIA standard set of General Conditions for buildings contains the following articles:

a. Contract Documents
b. Architect
c. Owner
d. Contractor
e. Subcontractors
f. Separate Contracts
g. Miscellaneous Provisions
h. Time
i. Payments and Completion
j. Protection of Persons and Property
k. Insurance
l. Changes in the Work
m. Uncovering and Correction of Work
n. Termination of the Contract

Such standard sets of General Conditions are complete documents, not blank forms. The user normally will not make any changes on the document itself but will expand or modify it through use of Special Conditions.

Special Conditions, Supplementary Conditions, or Special Provisions

This document adapts the General Conditions to the current project and can modify provisions of the General Conditions. It gets into such subjects as project location, special laws or regulations to be observed, handling of shop drawings, reporting requirements, scheduling requirements, permits to be obtained, wage scales, public relations, security, date for commencement, clean-up required, etc.

Appendix 4A (pages 66–71) is an extract of supplementary conditions for a cost-plus-fixed-fee with target cost contract for a steam-electric generating station. Close examination of this document will reveal many conditions and requirements with cost implications which the estimator must evaluate.

Specifications

There are two packages in the construction contract documents which describe the structures to be built. One package is the drawings whose primary purpose is to provide dimensional information on the project—location, size, relationship. The second package is the book of specifications which establish quality, performance, and methods to be employed.

There are three types of specifications that may be employed. The first is *proprietary* which expresses the design of an item by listing a specific make and model. A *descriptive* specification describes an item as to materials to be used, method of assembly or manufacture, features to be incorporated, etc. A *performance* specification describes an item only in terms of what it must do.

Standard descriptive specifications are available from professional and trade associations, such as the Construction Specifications Institute (CSI) or the American Concrete Institute (ACI). In addition, *reference standards* are available from these and other organizations such as the American National Standards Institute (ANSI) or the American Society for Testing and Materials (ASTM). A reference standard usually describes a procedure for accomplishing an action such as mixing or testing; it is not a specification in itself. However, it may be referenced in a specification and thus be incorporated by reference.

In the building industry, specifications are catalogued by Uniform Construction Index (UCI) *divisions*, as follows:

Division 1   General Requirements
Division 2   Site Work
Division 3   Concrete
Division 4   Masonry
Division 5   Metals
Division 6   Wood and Plastics
Division 7   Thermal and Moisture Protection
Division 8   Doors and Windows
Division 9   Finishes
Division 10   Specialties
Division 11   Equipment

Division 12   Furnishings
Division 13   Special Construction
Division 14   Conveying Systems
Division 15   Mechanical
Division 16   Electrical

Additional work breakdowns could be listed for other types of construction.

The subdivision of a division is a *section*. A section is the smallest biddable element of a project. It is complete within itself and can be withdrawn and given to a subcontractor or vender, if desired. A section will be organized along these lines.

Article 1.   General
  Description
  Quality control
  Submittals for approval
  Product, delivery, storage, handling
  Alternates
  Guarantees

Article 2.   Products
  Materials (reference standard)
  Mixes
  Fabrication
  Manufacturer

Article 3.   Execution
  Inspection
  Preparation
  Installation
  Application
  Performance
  Adjustment
  Cleaning
  Schedules

An estimator must carefully study the specifications to determine their impact on cost. A proprietary specification is essentially for a catalog item and price quotes may readily be obtained. Descriptive specifications may prove to be restrictive and costly since they tell the contractor exactly how to do something and this method may not be standard with the firm. Performance specifications are the most flexible and will allow selection of the most economical procedure for accomplishing an objective.

Design Drawings

Accompanying the book of specifications is a design drawing package. For buildings, these will be divided into architectural, structural, mechanical, and electrical sections. For other types of projects, other groupings may be appropriate. Drawings will describe structures dimensionally and will contain references to specifications within the specifications book. On relatively simple designs, specifications may be included on the drawings.

SPECIAL CONTRACT DOCUMENTS

Request for Proposal

The term Request for Proposal is often used synonymously with Invitation to Bid. A more general use is that associated with work which is yet to be designed and the client requests proposals from selected contractors to do the design and/or construction. The Request for Proposal (RFP) on design or turnkey work will define the project's scope, desired performance, location, and any other factors then known or desired by the client. The engineer responding to the RFP will then develop a conceptual design of the facility, provide an approximate estimate of its cost, provide an estimated schedule of construction and cash flow, and specifically respond to special requirements such as quality assurance/quality control programs to be used, cost control systems to be employed, identification of intended subcontractors, recommendations for procurement responsibilities for major components, etc. RFPs may be invited publicly or given only to selected contractors; however, the normal procedure is to send RFPs to only a few contractors with the capabilities to perform the work.

Proposal

The Proposal is the document furnished a client in response to an RFP. The term is often used to mean the same as Bid.

IMPACT OF CONTRACT DOCUMENTS ON COST ESTIMATING

Any contractor responding to an Invitation to Bid or Request for Proposal will expend considerable effort in preparing a response. In neither case will the contractor receive any compensation directly from the client for the costs associated with developing the response. Yet, these are costs that must be covered in some fashion. They become part of general overhead expenses (Chap. 13), a category which is a catch-all for expenses of the main office. These costs are distributed to all projects of the company as the general overhead markup. Thus, clients inevitably pay for the costs of bid and proposal preparation, assuming estimators properly include all main office costs in their estimates.

The extensive documentation associated with any contract reflects both client and contractor desires to protect their own interests. Such protection is not without cost. Each bond, each safety barrier, each schedule and each report required will cost the contractor money to provide; these costs are in addition to the direct labor, materials, equipment, and supervisory costs of the project. Thus, close scrutiny of contract documentation before responding is a must for contractors and their estimating groups. Stories are still being told of contractors first entering the nuclear power plant construction field who saw very general statements concerning quality assurance/quality control programs in the contract documents but passed them off as no real significance because they (the contractors) always did quality work. It was often too late when they found out that this requirement alone added 5 to 10 % to the total project cost.

Some of the contract document requirements are very specific and easy to cost. If a contractor has to provide a bid bond of $20,000 he or she can readily determine its cost from his or her surety. The dangerous requirements are the vague ones such as

Contractor will utilize a network scheduling system acceptable to the client.

This type of statement establishes a definite requirement (network scheduling system) but the contractor has no way of knowing how elaborate the system must be to be acceptable to the client—can it be manual and deal only with major systems or structures or must it be computerized and include work package detail complete with a family of reports? The difference is a great deal of money.

A common feature of a contract is the right of the client to retain a percentage of the earnings (about 10%) until the project is complete and has been accepted. This retention serves as leverage to insure that the contractor does not just walk off the job and say it is complete. However, this adds to contract cost. Each dollar retained is being denied the contractor for a period of time. With time-value of money, these dollars inflate in value. Realistically, a contractor bidding on a project should lay out the project income flow, identify the expected amount of retention for each period, inflate this amount to the end of the contract using time-value procedures, and account for the differences between inflated and original values as an added cost in the bid.

Liquidated damages may be allowed in some contracts. These permit a client to reduce a contractors earnings by a fixed amount for each day that a project extends beyond a set completion date. In theory, the amount set should be equal to an actual economic loss the client will suffer for each day of delay. If a contract contains such a provision the contractors must be extremely thorough in their planning to insure that the schedule can be met or, if there is some question, to add sufficient contingency to account for damages that will be assessed due to probable overrun.

SUMMARY

Contract documents are an essential feature of construction contracting. They describe the project and establish the rules of the game for its execution. They contain, often in legal language, a myriad of requirements to be met by both parties to the contract. Understanding of these requirements is as important as an understanding of the design because compliance will invariably result in expenditure of resources, time, and money.

REVIEW

Exercises

1. Appendix 4A is an extract taken from the Supplementary Conditions of a cost-plus-fixed-fee with target cost contract document package for a coal-fired steam-electric generating station. Assume you have been designated to head the team developing the response to this invitation to bid. Develop a list of those requirements contained in this extract that you feel deserve particular scrutiny by your team members in developing the target cost.

2. Using the unit-price bid example given in the text, calculate the amount that will be paid the contractor if measured quantities are as follows:

Item	Measured Quantity
Mobilization	1
Excavation, right-of-way	26,000
Excavation, borrow	13,500
Asphaltic concrete, in-place	14,500
Box culverts	2
Bridge	1

Questions in Review

1. Under what conditions is a lump-sum contract appropriate?
2. Why is a unit-price form of lump-sum contract used?
3. What is meant by lowest responsive bid?
4. In a competitive bidding situation with a unit-price contract to be awarded, how is the low bidder identified?
5. Under what situations is a cost-plus-fixed-fee contract appropriate? With target cost?
6. What is an escalation clause?
7. Explain the purpose of bid bonds, payment bonds, and performance bonds.
8. What is the relationship between General Conditions and Supplementary Conditions in a contract package?
9. Describe three types of specifications.
10. What is a reference standard?
11. What is a section of a specification division?

Discussion Questions

1. Should a client inviting bids or proposals pay the costs of preparing those proposals?
2. Is it fair for a client to expect fixed-price contracts in that the client or the client's engineer has designed the project, yet expects the contractor to assume all risk of building it?

APPENDIX

4A

—EXTRACTS—

SUPPLEMENTARY CONDITIONS
FOR COST-PLUS-FIXED-FEE WITH TARGET COST
CONTRACT

STEAM-ELECTRIC GENERATING STATION

* * * * * * * * * * * * * * * * * * * *

Field Overtime

CONTRACTOR shall be responsible for establishing the number of hours comprising his and his Subcontractors' basic work week, and shall submit such information with his Proposal. Both the straight-time costs and the premium portion or shift-differential costs of any overtime or shift work contained in the CONTRACTOR'S basic work week so established, or contained in any subsequent revision of the basic work week, shall be a part of Recoverable Costs, and the Bidder shall include any such costs in the Total Contract Target Price. Unless otherwise approved by the DISTRICT, the CONTRACTOR shall give the DISTRICT at least 72 hours advance notice of changes to the basic work week and at least eight (8) hours advance notice of overtime worked as a construction expedient (spot overtime), such as might be necessary to finish a concrete placement after normal working hours. Spot overtime shall be included in the Bidder's Proposal and shall not be the basis for any change in Total Contract Target Price.

If at any time any portion of the work is falling behind schedule, the CONTRACTOR shall furnish more men, work in double shifts or work overtime to maintain his schedule, all at no change in Total Contract Target Price, and the DISTRICT shall have the right to direct that such additional labor, shifts or overtime be worked to maintain the schedule at no change in Total Contract Target Price.

CONTRACTOR shall not be entitled to any adjustment in Total Contract Target Price for any shift work or for working his forces or Subcontractor's forces outside the basic work week; provided, however, if at any time during the progress of the work, the CONTRACTOR is not behind schedule and is instructed by the DISTRICT to work overtime or shift work, then this situation shall serve as a basis for a mutually agreed upon change in Total Contract Target Price.

Schedules

1. The successful Bidder (CONTRACTOR) shall commence work under this Contract upon notification of award, and shall schedule his operations so as to meet the following milestone dates and all other such Contract schedule completion dates tabulated in the DISTRICT'S Critical Path Method (CPM) schedule transmitted herewith:

Activity	Date
a. Commence Work	3/01/78
b. Begin Concrete	4/01/78
c. Raise Boiler Drum	11/17/78
d. Complete Steam Generator Building Structural Steel	3/05/79
e. Set Station Crane	5/03/79
f. Complete Station Structural Steel	7/16/79
g. Set Stator	9/06/79
h. Complete Coal Handling Facilities	4/04/80
i. Hydro the Boiler	7/04/80
j. Air test the Boiler	8/01/80
k. Complete Computer Installation	9/02/80
l. Complete Acid Cleaning	9/26/80
m. Turbine on Gear	10/03/80
n. Complete Blowout	10/24/80

o. Roll the Turbine with Steam   <u>12/05/80</u>

p. Commercial Operation   <u>5/01/81</u>

After notification of award, any changes in the DISTRICT'S CPM schedule will be subject to, and in accordance with, the provisions "PROGRESS OF THE JOB."

2. Prior to commencing field work, a detailed CPM Schedule shall be prepared by the CONTRACTOR based upon the milestone dates given in Paragraph 1 above and in the DISTRICT'S Critical Path Method (CPM) Schedule referenced herein.
3. The CONTRACTOR'S CPM Schedule shall, in addition to all requirements specified above, as a minimum, include the following specific data and supplemental information:
    a. Further schedule breakdown showing the expected progress of the work to be performed in completing each scheduled CPM activity.
    b. Description of the scope of each breakdown activity.
    c. Man-hours required to complete each breakdown activity.
4. The CONTRACTOR shall familiarize himself with equipment and material delivery dates and consider them when planning his work. CONTRACTOR-proposed changes to these schedules must be submitted to the DISTRICT, who shall have final authority as to whether or not such changes can be effected.

Access and Work Area

1. Access to the site via the county road and Road "B" might be hindered during construction of the railroad. Bidders are advised that alternate access routes during this time must be approved by the DISTRICT.
2. The main plant area has been fenced. The CONTRACTOR shall be responsible to maintain his forces within the fenced-in area or, when working outside this area, to establish temporary work areas acceptable to the DISTRICT and identify these areas with temporary fencing, stakes, or other visible means.

Progress of the Job

1. Schedule
Within thirty (30) days from date of notification of award of Contract or after the dates for performance are fixed, whichever is later, the CONTRACTOR shall submit a progress schedule showing the order in which he proposes to carry on the work, with dates when

he will start the several parts of the work and dates of completion of
each of the several parts of the work. Such Progress Schedule shall
in all respects conform to the requirements presented in . . . of this
Contract Document. When approved by the DISTRICT, this Progress
Schedule shall be subject to the provisions of . . . .

2. *Reports*

In compliance with the request of the DISTRICT, the CONTRACTOR
shall promptly provide reports on his undertakings under this Contract without any change in the Total Contract Target Price.

3. *Time for Completion*

The Contract will be awarded to the CONTRACTOR in reliance
upon his maintaining the completion, delivery, drawing, and submit-
tal date(s) established in the Contract Documents, and the CON-
TRACTOR shall make provisions to effectively proceed with the
work immediately upon notification of award to him.

The CONTRACTOR expressly covenants and agrees that each part of
the work shall be completed in accordance with . . . , and that all parts
of the work under this Contract shall be completed on or before the
date(s) established in the Contract Documents. The CONTRACTOR
agrees that in making this undertaking and scheduling the work involved,
he has taken into consideration, made allowances for, and will provide
corrective action for all of the ordinary delays and hindrances incident
to such work whether growing out of delays due to normally inclement
weather, in securing materials or workmen, or otherwise. In the event
that the CONTRACTOR is delayed in the performance of the work as
a result of causes beyond his control and which he could not have reasonably anticipated and without his fault or negligence such as acts of
God, fire, flood, war, governmental or judicial action, or is delayed by
strikes or work stoppages which are a result of causes beyond his control and without his fault or negligence, the time specified in the Contract Documents for completion of the work will be extended for performance of the work, provided the CONTRACTOR gives written notice
thereof to the DISTRICT or ENGINEER within fifteen (15) days after
the commencement of such delay. If the CONTRACTOR encounters
extra costs as a result of delays which are beyond his control and which
he could not have reasonably anticipated and without his fault or negligence, including those delays which are due to the actions of the DIS-
TRICT, but not including delays resulting from unavailability of labor,
strikes, work stoppages, or slowdowns or any labor disputes, or delayed shipment of CONTRACTOR-furnished material or equipment,
the CONTRACTOR shall promptly give the DISTRICT notice, in writing,
of such extra costs to which he believes he is entitled and he shall furnish
satisfactory documentation of such extra costs. The CONTRACTOR

shall take immediate action to minimize the effect of such extra costs. The CONTRACTOR shall recover no such extra cost if incurred more than 15 days prior to giving the notice herein required.

Any adjustment of the Total Contract Target Price and any additional time to which the CONTRACTOR is entitled shall be mutually determined by the CONTRACTOR and the DISTRICT, and the CONTRACTOR shall be notified of such adjustment in writing by the DISTRICT. If the CONTRACTOR and the DISTRICT fail to agree upon such adjustments, the DISTRICT shall determine an equitable adjustment therefor and modify the Total Contract Target Price and the Contract time for completion, accordingly.

Schedule and Progress Reports

In addition to any other requirements in this Contract Document, the CONTRACTOR shall furnish to the DISTRICT, thirty (30) days after notification of award of the contract, a Critical Path Schedule of expected progress for the work to be performed under this contract, conforming in all respects to the requirements of . . . .

Any data required by the CONTRACTOR from other contractors, or the DISTRICT, shall be indicated merely by an arrowhead activity. Contractor's assumed times for these activities will be subject to approval of the DISTRICT. Such schedules as furnished by the CONTRACTOR shall be accompanied by appropriate computer runoff sheets to show as a minimum the ESD (Earliest Starting Dates) and the list of total float in ascending order.

The CONTRACTOR shall submit to the DISTRICT, his purchase order lists showing his various suppliers, purchase order numbers, date, a description of the material involved and the delivery date specified. The CONTRACTOR shall also submit copies of subcontracts which he executes showing date, a description of work or material to be furnished and delivery dates. Such information shall be provided promptly so that the DISTRICT will be aware of the progress and adherence to the schedule by the CONTRACTOR in placing of orders and adherence to the specification requirements of the Contract Documents.

During the course of the work, the CONTRACTOR shall regularly update his CPM schedule for the current aspects of his work and officially submit this updated information to the DISTRICT on a monthly basis on the last day of the month. If the DISTRICT decides that the CONTRACTOR'S original CPM schedule has been sufficiently changed in the course of his work, the CONTRACTOR shall submit a new and revised CPM schedule within 15 days of such notification.

The CONTRACTOR shall confer on a regular schedule basis with the DISTRICT and with the other CONTRACTORS for the purpose of formulating the detailed work in accordance with the schedules and coordinating his work with the work of other CONTRACTORS.

The CONTRACTOR shall conform to the DISTRICT'S overall program to achieve the rapid completion of the project as a whole and within the limits of the agreed upon schedule which form part of this contract.

The CONTRACTOR shall have his trades afford all other trades under the control of other contractors every reasonable opportunity for the installation of their work as well as for the storage of their material and equipment.

When the CONTRACTOR is required to place, install, or connect up material or equipment furnished by others, the CONTRACTOR shall notify the DISTRICT in writing when such equipment or material will be needed, and the CONTRACTOR shall cooperate with the DISTRICT in arriving at the best workable overall scheduling of such work.

The CONTRACTOR shall provide the DISTRICT with a schedule of submittals of drawings for approval, delivery of material or equipment, and purchase order numbers, dates, descriptions of material involved and specific delivery dates.

* * * * * * * * * * * * * * * * * * * *

PROJECT CONTROL
THROUGH INTEGRATED
COST ENGINEERING

5

INTRODUCTION

The first four chapters were designed to describe the estimating challenge and introduce you to the world of contract construction. In this chapter the objective is to describe the total system of project control so that you can better understand the role of estimating. Following chapters will concentrate on cost components and the actual estimating processing.

Project control is achieved through the application of cost engineering. Since this term does not enjoy a common definition throughout the construction world it is appropriate at this point to give its definition as used in this text and as officially defined by the American Association of Cost Engineers:

Cost engineering is that area of engineering practice where engineering judgement and experience are utilized in the application of scientific principles and techniques to problems of cost estimation, cost control, business planning, and management science.

As such, cost engineering encompasses the functions of planning, estimating, scheduling, and cost control in construction.

For a cost-engineering system to be effective in project control, the system must be activated from the time bidding documents are first received by a contractor and maintained until the project is closed and historical records completed. The key to an effective system is the integration of all functions through use of a common work package breakdown structure and a well-designed code of accounts plus the maintenance of a complete and valid data base.

WORK PACKAGE BREAKDOWN STRUCTURE (WPBS)

In scheduling a project, the scheduler learns to identify activities. These activities are logical pieces of the total project that are easy to identify with a start and a completion. Following identification of activities they are arranged in a logical fashion for accomplishment by the responsible agency or crew. Work packages are the same as the activities used in scheduling, the term *work package* being preferred because it better suits the intended meaning which is

A work package is a well-defined scope of work that usually terminates in a deliverable product. Each package may vary in size, but it must be a measurable and controllable unit of work to be performed. It also must be identifiable in a numerical accounting system in order to permit capture of both budgeted and actual performance information. A work package is a cost center.

The theory behind project control through work packaging is simple—to manage a whole operation, you manage and control its parts, these parts being work packages in the case of construction. A project is planned by work packages, estimated by work packages, scheduled by work packages, and cost controlled by work packages.

The size of a work package can vary as is reflected in the definition. However, each package must be easily defined in terms of total content and have a beginning and an end. There can be a number of levels of work packages. For example, a total building as part of a larger complex is a work package. Yet, it can be broken down into successively more detailed work packages as illustrated in Fig. 5-1. All but the last level serve as locators of work and there can be as many levels of these locators as required for the type of project. In this breakdown, Level II contains the main building components. In Fig. 5-1, Level III breaks these down for foundation construction. Level IV is the lowest level of location detail. Figure 5-1 shows this level for the pile cap work. Finally, Level V identifies the specific crew tasks at the location previously identified (resteel of pile cap of building foundation in Fig. 5-1). At each level, it is easy to visualize the entire package in terms of physical appearance upon completion or as an action that has a beginning and an end. For purposes of comparison, assume that the building had been catalogued by level, such as subsurface, first floor, second floor, roof, grounds, etc. Such a division of work would create an impossible control situation. Where is the boundary between the first and second floor? How is the air conditioning system divided among floors? Such questions would continually arise because the work is not catalogued in a manner compatible with the way work will be done.

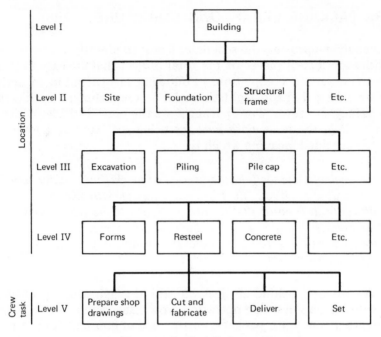

Fig. 5-1 Example of a Work Package structure.

To complete a work package requires the expenditure of the resources of construction—labor, installed equipment, materials, and construction equipment. Time also will be required to complete the package and the package will have a total cost. The total control system involves first the isolation of component work packages; second, the estimation of resource requirements, time and cost for each work package; and finally, control of construction by relating actual resource and time consumption with that originally estimated.

CODE OF ACCOUNTS

In ordering materials from a catalog, each line item ordered will have a catalog number which is a series of numbers and/or letters. This number is part of an overall code of accounts used by the company. That number usually contains several items of information such as department, manufacturer, and specific item identification. It uniquely identifies the item ordered and contains no perishable information such as cost, order–ship time, or storage bin location. A code of accounts is also appropriate to construction since it provides the common language of identification to be used by all cost engineers during project control.

Design of Code of Accounts

Basic Requirements.   A well-designed code of accounts will satisfy the requirements listed below. Each of these requirements will be discussed in detail in following paragraphs.

1. Must catalog all needed information
2. Must facilitate consolidation of information at levels of detail from very broad to very detailed
3. Must allow for extraction, sorting, and summary of selected information from within the total
4. Must be compatible with other accounting systems used in the company
5. Must be suitable for computerized processing
6. Must be usable by typical operating personnel
7. Must support both current and historical needs

Cataloging of Construction.   The code of accounts must be designed to provide the information needed by its users. If a cost is reported, is it a direct cost, indirect cost, or otherwise? To what work package does it apply and is it an overall cost, labor cost, materials cost or other? If a production unit is reported, to what resource does it apply? In general, a contractor will find a code of accounts structure similar to that illustrated in Fig. 5-2 to be useful. In this sample a total of 16 letters or numbers represented by Xs are used. The dashes and periods are not actually part of the code number; they are shown to visually separate groups and subgroups. These 16 Xs are further arranged into four groups. The first group is a single digit representing accounting classification. If letters are used, 26 classifications are possible; if numbers are used, 10 are possible if zero is included. The second group is six digits, the first two digits being used to identify the major structure/system, the next three the substructures or subsystems and the last being to uniquely identify work packages that are otherwise identical (such as multiple foundation pads). A contractor specializing in standard types of construction will use the same numbering system from project to project. Possible codings for the first five letters within this group are

02.000	Boiler Plant
02.200	Boiler Plant Pressure Components
02.210	Boiler Plant Waterwalls
02.220	Boiler Plant Superheater
02.221	Boiler Plant Superheater Header
02.222	Boiler Plant Superheater Crossover Piping

. . .

16.000   Circulating Water System
16.100   Circulating Water System Intake Structure
16.200   Circulating Water System Chlorination System

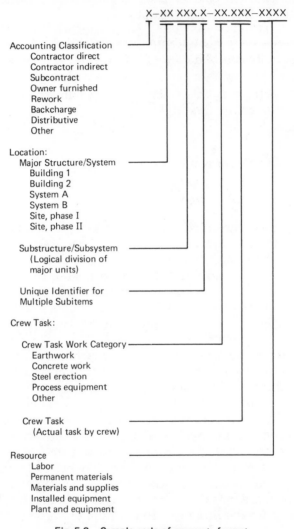

X–XX XXX.X–XX.XXX–XXXX

Accounting Classification
    Contractor direct
    Contractor indirect
    Subcontract
    Owner furnished
    Rework
    Backcharge
    Distributive
    Other

Location:
    Major Structure/System
        Building 1
        Building 2
        System A
        System B
        Site, phase I
        Site, phase II

Substructure/Subsystem
    (Logical division of
    major units)

Unique Identifier for
Multiple Subitems

Crew Task:

    Crew Task Work Category
        Earthwork
        Concrete work
        Steel erection
        Process equipment
        Other

    Crew Task
    (Actual task by crew)

Resource
    Labor
    Permanent materials
    Materials and supplies
    Installed equipment
    Plant and equipment

Fig. 5-2   Sample code of accounts format.

Note how the descriptions go from broad to detailed. The coding 02.000 refers to the total boiler plant so it is a summary level. As individual components within the boiler plant are identified, additional numbers replace the zeros to the right. An example of greatest detail is that of the coding 02.222 which represents the crossover piping of the super-heater of the boiler plant. Note also that this group of digits is a locator of work and not the crew task itself. As a matter of good coding format, the first two digits should have exactly the same meaning from project to project. The last three digits may be reused as the first two digits change. In the example, 200 in the last three positions means boiler pressure components when used in conjunction with 02 in the first two positions while 200 means chlorination system when used with 16 in the first two positions. However, when the five digits are considered as a group, the same coding will always have the same meaning.

Work package crew task coding is the third group and follows the same logic as for structures and systems. Possible codings in this group are

01.000   Site Preparation

. . .

02.000   Earthwork

. . .

03.000   Concrete Construction
03.100   Slabs on Grade
03.110   Formwork, Slabs on Grade
03.120   Resteel, Slabs on Grade
03.130   Concrete Placement, Slabs on Grade
03.131   Concrete Placement, Direct Chute, Slabs on Grade
03.132   Concrete Placement, Crane and Bucket, Slabs on Grade
03.133   Concrete Placement, Pumping, Slabs on Grade
03.134   Concrete Placement, Conveyors, Slabs on Grade
03.200   Elevated Slabs
03.210   Formwork, Elevated Slabs

Here again notice how summary level coding is accomplished through use of zeros to the right while detail is provided by use of other numbers. Consistency is maintained in the use of numbers in the last and next to last places. The number 1 in the next to last place is reserved for forming, 2 is for resteel, and 3 for concrete for all concrete construction coding (they would have other meanings under site preparation, earthwork, etc.). In the last place, the number 1 is reserved for placement by direct chute, 2 signifies crane and bucket, etc. when used in conjunction with a 3 in the next to last place.

The final grouping is for resource coding. Four digits have been allowed but more or less can be used. Normal practice is to reserve groups of digits for categories. For example, numbers from 0001 through 0999 may be reserved for labor, 1001 through 1999 for permanent materials, etc. Here again a pyramidal coding system is possible. Examples include

0100	Job Overhead Personnel
0110	Construction Management Personnel
0111	Construction Manager
0112	General Superintendent

. . .

0200	Craft Personnel
0210	Carpenters
0211	Carpenter Foreman
0212	Carpenter Journeyman
0213	Carpenter Helper

The number of digits or letters used and the order of groups within the code are at the discretion of the contractor. If excessive detail is used, the system may be unmanageable. On the other hand, too basic a system will not allow for future growth of the firm.

Consolidation of Information at Various Levels of Detail. With the pyramidal system of coding as described above, it is possible to get both summary and detailed information. For example, if employee-hours are reported against a coding of 02.222, Superheater Crossover Piping, this information can automatically be credited simultaneously to three summary levels above it—02.220, Superheater; 02.200, Boiler Pressure Components; and 02.000 Boiler Plant. In other words, data is accumulated at a number of levels allowing summary information to be developed as well as detailed information.

Extraction and Sorting Potential. This capability is an extension of the summary capability just discussed. By assigning standardized meanings to each group of digits, it is possible to mix and match data as needed through selective withdrawal. Whereas the summary capability allows summarization of data within a group, the extraction and sorting potential allows it among groups. For example, if concrete slabs on grade always have the same coding, it is possible to summarize production, employee-hours, or cost of all slabs on grade throughout the project simply by selecting all reports that carry the coding for slabs on grade in the work package identifier. Another possibility is the summary of all employee-hours expended to date in a major structure or the entire project.

Compatability with Other Company Accounting Systems. As a matter of economics and efficiency, duplication of effort must be avoided in accounting and control whenever possible. If one division of a company uses one code of accounts and other divisions use different codes, it is difficult to share information or to move personnel from project to project without retraining.

Suitability for Computerized Processing. When using codes of accounts as complex as those proposed, it is usually assumed that the system will be computerized. We take advantage of the ability of a computer to input, process, and store vast quantities of information and to selectively retrieve it and output it into virtually any form desired. The capabilities and relatively minor cost of micro- and minicomputers make computerized project control possible for even small contractors. Commercial time-sharing is another excellent option for all contractors particularly because it makes available many existing software programs suitable for construction project control.

Usability by Personnel. The designers of a coding system must keep in mind the personnel who will be using it. If a complex coding system is used, it is essential that professional cost engineers maintain basic control over all coding operations. A contractor cannot expect crew foremen to become proficient in code of accounts use so an individual trained in cost engineering must be available to receive raw information from the crews and properly code it for processing into the reporting system. A small contractor with no professional cost engineering staff probably will find it necessary to utilize a very limited coding system.

Use for Current and Historical Needs. The code of accounts permits organization of all planning, estimating, and scheduling work. During construction it permits comparison of actual to budgeted progress using a variety of reports. When a project is completed, it is essential that appropriate summary information be placed in the historical files, again by proper code. This data can be quickly extracted as needed for assistance in cost engineering of new work.

Use of Zeros in Coding System

In the examples given above, the use of zeros to the right of the last nonzero digit indicated summary level information. The use of all zeros in a given group means that that group is not being used (that is, not applicable) in the identification of the item being reported. For example, the employee-hours of the construction manager (a job indirect item) are being reported, the coding might be as shown in Fig. 5-3.

Code of Accounts Applications

The code of accounts is used throughout the life of a project. Figure 5-4 illustrates the many places the work package identifier will

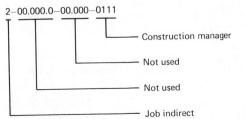

Fig. 5-3 Sample coding.

find use. For all other reporting and accounting purposes the code permits ready compilation and outputting of information that would be impossible to produce by longhand methods on a large project. The basic code structure we have discussed is the standardized portion that would be described in detail in a catalog available to all users. The basic structure can be augmented or expanded by the addition of digits either before or after the basic portion. For example, two or three digits preceding the basic code may be used for a project number or for entry of a key number which converts an item of information to another reporting system used by a client or by a regulatory agency to whom specialized input may be required. More than 20 numbers (or letters) may be employed with larger computer systems.

Standardization of Units Reported

The code as described is only an address for a piece of information being reported. That information can be cost, employee-hours, production units, or some other quantity. To avoid any misunderstandings on what is being reported, it is essential that a contractor's code of accounts catalog include standard units for reporting of all resources. For example, craft labor will be reported as employee-hours, concrete as cubic yards, structural steel as tons, construction equipment as equipment-hours, etc. Units selected should be those which are used in the estimating stage since the objective is to generate reports which permit direct comparison of actual performance to estimated. Subprograms within the computer program can convert basic units of measure to cost as necessary.

BID DEVELOPMENT FOR A MAJOR PROJECT

The discussion on work packaging and codes of accounts was for the purpose of better understanding the total bid preparation process and later project control if the contractor is the successful bidder. Discussion now proceeds to the steps involved in both construct-only and turnkey construction. In this discussion the term *project team* will be used quite frequently. The project team is composed of those individuals

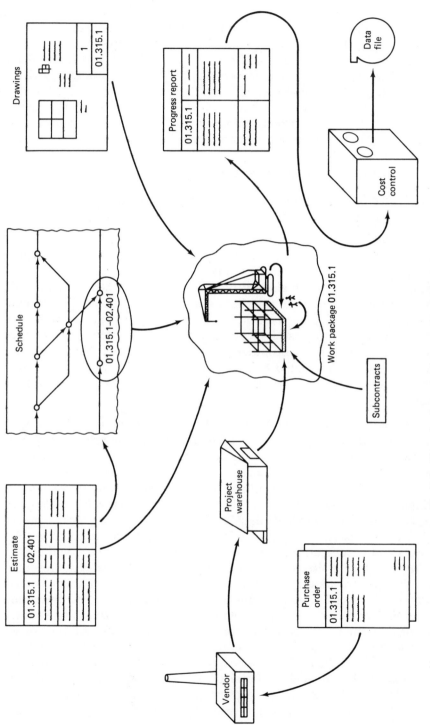

Fig. 5-4 The role of a code of accounts in project control.

Drawings

01.315.1

Progress report

01.315.1

Data file

Cost control

Schedule

01.315.1–02.401

Work package 01.315.1

Subcontracts

Estimate

01.315.1	02.401	

Project warehouse

Purchase order

01.315.1

Vendor

in charge of major functions that will come into play in bid development and total project control. The project team is a task force that is specially formed for a specific mission. The head of the team is often called a project manager and other members will come from estimating, scheduling, cost control, procurement, and other affected staff divisions. Assignment to a project team is temporary, its members are still organizationally a part of their original sections. In fact, an individual may simultaneously be a member of several project teams.

Bid Development for Construct-Only Projects

Figure 5-5 is a graphic representation of the steps to be described. It should be consulted to follow the discussion.

Initial Management Action. Contractors will not necessarily spend the time and effort required to develop a bid for all projects for which an invitation has been received. On the contrary, they may pass an invitation to bid if they are already fully committed on other work, if the project appears to contain unusual risks, if the designs are poorly executed, or for some other reason. Thus, the first step in the process is listed as a management action to review the project, determine whether a bid will be prepared, and otherwise develop guidance for the project team if they are instructed to go ahead with the bid. Management will certainly be assisted in this review by potential members of the project team.

Initial Project Team Action. With a decision to bid, a project team is formally designated. As their first action the project team will review the contract documents in detail to insure that they have a full understanding of project requirements. They particularly will look for the unusual in terms of deadlines, nonstandard types of construction, unfavorable site conditions, and similar features which will tend to drive up the project duration and cost. Visits to the site are essential to relate the documentation to the project since what may appear to be routine designs and specifications may become nightmares of work and cost in reality due to the site peculiarities. They also will discuss the resource situation (labor, materials, and equipment) in the project area to determine what special investigations might be appropriate to determine availability and cost. The planners next "build the project on paper." Having determined all requirements of the proposed contract, they break them into work packages. The level of work package detail depends on the situation. If the design is conceptual only, work package breakdown is necessarily broad, such as Level II of Fig. 5-1. If the plans are complete the use of more detailed packaging can be considered (such as Levels III, IV, or V of Fig. 5-1). As an objective, the level of work pack-

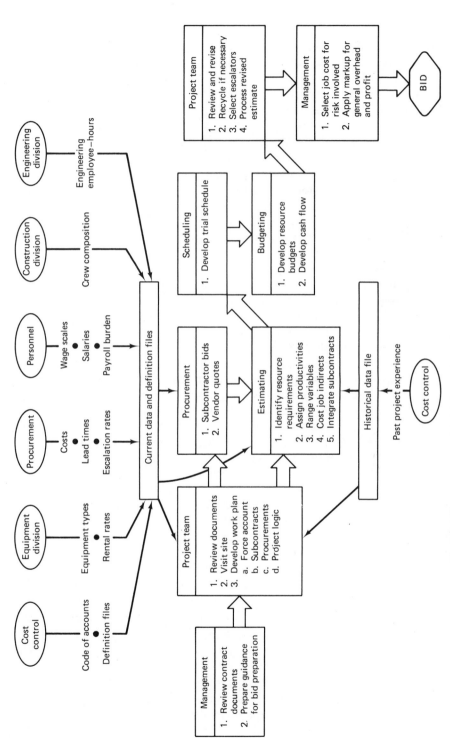

Fig. 5-5 Flowchart for bid development for a major construct-only project.

age detail should be that which permits isolation of subcontract packages from in-house packages, since this is a major planning action. Following work package definition is the outlining of project logic, preferably in the form of a logic network. This network contains no time factors but does develop the Work Plan by showing interrelationships and dependencies of the various work packages. It is the skeleton upon which a target schedule is later developed after durations are determined during the estimating function. Thus, in its final form a Work Plan provides a network outline of the way the project will be designed and/or constructed, an identification of all force-account work packages, and an identification of all subcontract packages. In addition, if appropriate for the project involved, the Work Plan can include a summary of major items of installed equipment or other components which have significant procurement lead times.

Estimating.   In estimating, the ultimate objectives are to identify resource requirements, costs, and durations. To do this, the estimator must quantify permanent materials and installed equipment from the drawings, determine other materials and supplies required for placement of permanent materials and installed equipment, select appropriate crews and construction equipment to perform the work, and make appropriate assumptions of expected productivity from these crews and equipment. If plans are relatively well-defined, permanent material and installed equipment quantities are determined by measurement or calculation from the drawings, this process being either completely manual or computer-assisted. If plans are conceptual only, estimates of quantities must be based upon previous experience on similar work and estimator judgment. Ideally, the company has had experience with similar projects and can go to the historical data file to obtain quantity ranges. Lacking this, they may be able to look to the experience of others in the field or, lacking any direct source, must combine the experienced judgment of the staff to estimate quantities. Crew compositions will tend to be standardized within a company, so base data on these crews—their composition and productivity rates—will be found in the definition and historical data files. These will be supplemented by data from the current data file to include wage rates, payroll burden, and anticipated escalation. In approaching an estimate the estimator may divide a work package into its component resource requirements and then cost each resource separately. Alternatively, he or she may use a factoring approach which starts with a base unit (major item of installed equipment or major resource quantity) and then, using appropriate multipliers (factors), converts this base unit into employee-hours, tons of resteel, supporting equipment cost, or other resource quantity or cost. Parameters provide another approach to approximate estimating. A parameter relates the cost or quantity of each work package to some basic dimension of the total structure or system. For housing, the quantities of resource quantities included in work packages may be related to square footage or volume of living space (such as board feet of framing per square foot). For power plants, the relationship can be resource quantity per kilowatt of output; for process plants it can be related to plant capacity in barrels, tons, or other units of measure. Whatever the approach, the estimator uses the experience of past projects as recorded in the historical data files in making an estimation.

In practice, a combination of detailed, factor, and parameter estimating may be employed on a single complex project. Major constructed items may be costed on a detailed basis while packages such as quality control or utilities costs are handled on a parameter or factor basis.

In making estimates, the estimator may choose to develop an estimate as if the project were being built completely at the present time and in the base area since the best historical data will be from the base area. The base area data is then extended to the area of the project and to the proper time frame by making appropriate adjustments to rates and unit prices. This approach permits an overall comparison of construction under base conditions to that under adjusted conditions. Such a comparison is often useful during overall review of a project since an estimator may over- or underadjust overall and this condition may not be suspected without the overview comparison. As another approach, the estimator may introduce adjustments in each work package as the estimate proceeds. Perhaps the ultimate system is for an estimator to enter units and rates applicable to the project area at time zero, enter expected inflationary rates for resource costs, and also employ probability by expressing variables in terms of low, high, and target. This system must be computer-based. Such an approach has two major advantages. First, the estimator no longer has to come up with a single figure for each variable—he or she deals in ranges which is more realistic. Second, at the estimating stage, the timing of work packages is still unknown and the estimator is not forced to assume a time for application of inflation factors. This will be taken care of by the computer since it can simultaneously schedule and apply inflation factors. Such a system will be discussed in a later chapter.

Another estimating task is the costing of job indirects. The estimator must identify all project overhead personnel, the time they will be on the project, requirements for offices, warehouses and other support structures, utilities, etc. Check lists for these will be provided in a later chapter.

Still another estimating task is the incorporation of subcontractor bids and vendor quotes obtained from the procurement group into the estimate.

Procurement.   The two major functions of procurement are purchasing and subcontracting. In large construction organizations, a separate procurement staff group is established to handle these functions because of the legal intricacies of both actions. Requirements are furnished to the procurement group by engineering, construction, and other groups, but it is the procurement group that must develop the appropriate contract documents and handle all bidding or negotiations. From the planning function will come requirements for subcontracts and major procurements to be included in a bid. Procurement will be furnished copies of appropriate plans and specifications upon which to solicit subcontractor bids and vendor quotations. These bids and quotes, in turn, will be turned over to the estimating group for incorporation in the overall estimate. For some items, exact quotes may be impossible to obtain because of design uncertainties or limited time for bid preparation. In such cases, procurement will provide current costs (actual or estimated) and expected annual escalation.

Scheduling.   Scheduling is really the combining of a logic diagram (as determined during the planning function) with the durations (which result from the estimating function). The first trial schedule is based upon durations derived from estimating. It is unlikely that this trial schedule will prove to be entirely satisfactory in terms of meeting project milestones or in providing an acceptable leveling of resources, so adjustments may be necessary in the handling of the project and its various work packages. These adjustments may involve changing crews assigned or work methods employed, revising decisions on subcontracting, or resequencing of the overall project.

Budgeting.   Budgeting is the planning of resource requirements against time. It combines the output of scheduling and estimating. It is part of project planning to schedule labor and equipment so that on-site requirements do not go through severe ups and downs. This is called resource leveling. Ideally, the labor force will build up smoothly, peak, and then demobilize slowly so hiring, cross-training, reassignment, and discharge programs can be smoothly accomplished. If this is not done, there will be abnormal personnel turnover with its resultant detrimental effect on production efficiency. In the case of equipment, poor planning of its use among the various work packages can result in excessive numbers of equipment items on the job and significant amounts of nonproductive time. As mentioned earlier, resource leveling may require recycling of portions of the estimate. Budgeting is facilitated by work packaging since each work package is identified on the schedule as to timing and all incorporated resources can be readily listed.

Project Team Action.   The project team should be available at any time to resolve problems not within the ability of the divisions to handle among themselves. They will have a very definite review role once all components of the target estimate, trial schedule, and budgets have been developed. In effect, they give it a last team look, make any adjustments required, and prepare to defend the estimate to management.

Management Action. In a meeting with senior management, the project team has the task of defending their estimate. Having gone through the process described, they can assure management that it is based upon the most reliable information available concerning actual company experience and known factors. From the data presented, management can select a job cost which incorporates an acceptable degree of risk through addition of a contingency add-on. Management also will, at this time, apply markup factors for general overhead, time-value cost of retained progress payments, and profit. Following this, the final bid can be totalled.

Supporting Actions. Referring again to Fig. 5-5, we can note that the flow of actions described above has been dependent upon a series of data files labeled as current, definition, and historical data files. The current file contains perishable information such as wage rates and equipment rental charges. The definition file contains information on standard crews, standard parameter measures, etc. The historical file contains production and other data from past projects. Collectively, these files provide the raw information needed by the project team and their staff in the execution of their respective functions.

The Turnkey Project

The major differences between a turnkey project and a typical construct-only project are the addition of design to the project and the lack of detailed plans and specifications. Following are the basic steps involved.

1. A project team is appointed.
2. The team carefully analyzes the documentation from the client to identify all requirements and desired features of the project.
3. Based on client requirements, a conceptual design outline for the facility is prepared which lists major plant components, supporting transportation network, and any other features which require further design definition.
4. The conceptual design outline is broken into components and these components are assigned to various design engineers on the staff who will further expand them for the purpose of defining permanent structures and installed equipment.
5. The conceptual designs of the design engineers are reviewed by the

project team both for engineering adequacy and constructibility. The product of this review is a general construction plan for determination of construction costs by the estimators.

6. The cost estimators estimate the costs and durations of each of the component work packages in the design. They also evaluate, in coordination with construction engineers, the requirements for construction support facilities and their cost.

7. Schedulers convert the design and construction plans into schedules for design and construction. These schedules are Level I or II schedules meaning that they deal with major systems or structures, or the major subsystems or substructures of which they are comprised. More detailed schedules will be developed later in the design and construction sequence if the contract is awarded.

8. Cost engineers develop a cost control program to be implemented upon receipt of the contract. It will use the schedule as a base and contain field reporting, costing, and data storage features so that reports can be generated for each level of management.

9. The senior team members review the total input with contributing engineers to insure that all information is present which is needed to completely respond to the client's request for a proposal. They, in turn, will present the total package to senior management in the firm where it will be further reviewed. At this level, a decision is made on an appropriate markup to reflect costs of overhead, profit, and contingencies.

10. The proposal is submitted to the client. It will contain at least the following:

   a. A conceptual design.
   b. An engineering schedule.
   c. An estimated design cost.
   d. A construction schedule.
   e. An estimated construction cost.
   f. A listing of major installed equipment with proposed procurement responsibility (owner or contractor).
   g. A listing of all services to be performed.
   h. A description of the cost control system to be employed.

PROJECT CONTROL DURING CONSTRUCTION

Among the products of estimating are resource quantities and costs. If the work packaging approach described has been used, then these quantities can be isolated by work package. If the contractor wins the contract the project control initiated during the bidding cycle will be continued. The estimated quantities in each work package now become the target or budgeted quantities since the bid was based directly upon them. By setting up a control system during construction which focuses on work packages, the contractor will be able to directly compare actual expenditures of resources and money to that budgeted.

The Control System

For each work package there is a total amount of effort required which is measured in terms of tons of steel, cubic yards of concrete, square feet of surface, or a comparable quantity. Actual progress on an individual work package can then be readily measured by directly comparing units placed to date to latest estimate of total units required.

$$\text{Percent complete, work package} = \frac{\text{Units placed to date}}{\text{Total units to place}}\,(100\%)$$

A problem arises when a contractor seeks to determine an overall percentage complete on a total project which is a mix of earthwork, steelwork, concrete work, etc. To calculate overall percentage complete, the contractor refers back to individual work packages and identifies the employee-hours budgeted for each package. By dividing the budgeted employee-hours by the latest estimate of total units the contractor gets a figure which is budgeted employee-hours per production unit. Since virtually every work package has employee-hours of effort associated with it, an employee-hour is a logical weighting unit of measure that will permit calculation of overall percentage completion figures. However, to do this a new term is introduced: earned employee-hours.

Earned Employee-Hours (EEH): The value obtained by multiplying the budgeted employee-hours per production unit by the actual number of production units placed.

Its use is best explained by the example report shown in Table 5-1. The data in this table relates to a small reinforced concrete project composed of three work packages—forming, resteel placement, and concrete placement. The latest quantity estimates are 1000 SF of forms, 0.3 tons of resteel, and 25 CY of concrete. In the original estimate, employee-hours were budgeted for each package as shown in column (3). Column (4) is derived by dividing budgeted employee-hours in column (3) by total quantity from column (2). These are the budgeted amounts against which control will be based. The next two columns show the amount of work scheduled for completion to date expressed in both production units and EH. You will note that the EH in column (6) are equal to the EH/unit from column (4) multiplied by the target quantity in column

TABLE 5-1. Sample Project Report Information

(1)	Latest quantity estimate (2)	Employee hours Budgeted EH (3)	Employee hours EH Unit (4)	Target to-date Quantity (5)	Target to-date EH (6)	Actual placed Quantity (7)	Earned EH (8)	Percent comp. (9)	Actual EH (10)
Forming	1000 SF	100	0.1	1000 SF	100	1000 SF	100	100%	120
Resteel	0.3 tons	6	20	0.3 tons	6	0.2 tons	4	67%	4
Concrete	25 CY	50	2	5 CY	10	0	0	0%	0
		156			116		104		124

$$\text{Overall \% complete} = \frac{\text{Earned employee hours}}{\text{Total budgeted EH}} = \frac{104}{156} = 67\%$$

(5). The next three columns relate to the latest progress report. It shows that 1000 SF of forms have been placed as well as 0.2 tons of steel. No concrete has been placed. Using these values, the contractor multiplies each by the corresponding value in column (4) to get the EEH value of column (8). In the case of forms, the contractor has placed 1000 SF. The budget allowed 0.1 EH/SF so the contractor has earned 100 EH in doing this work (100 EEH = 0.1 EH/SF × 1000 SF). The workers have placed 0.2 tons of resteel and so has earned 4 EH (4 EEH = 20 EH/ton × 0.2 tons). No concrete was placed so no hours were earned. Earned employee-hours are entered in column (8). Column (9) provides percent complete for each work package and is found by dividing column (8) by column (3) or column (7) by column (2) since they are directly related. In this case we see that forms are 100% complete, resteel 67% complete, and concrete is 0% complete. We now want to find overall percent complete. This is done by adding up the budgeted EH in column (3) and the EEH in column (8) and then dividing the total from column (8) by that from column (3). The result is overall percentage complete which is 67%. To continue with the control system, note column (10) which records actual employee-hours. Here we see that the forming crew took 120 EH to complete a job estimated at 100 EH. However, this package is at least complete as planned. For resteel, column (5) tells us that the contractor had planned to have 0.3 tons in place and to be 100% complete. However, only 0.2 tons were placed earning 4 EEH and actual employee-hours were also 4. This would indicate that the contractor is behind schedule on this package but not expending more EH/unit than planned. Since no concrete has been poured there are no actual or earned employee-hours to record. However, note in columns (5) and (6) that some work had been planned for completion so this item is behind schedule also. We now use the employee-hour totals from columns (6), (8), and (10) to give us more information on this work package.

Employee-hours scheduled to date = 116

Employee-hours earned to date   = 104

Employee-hours actual to date   = 124

If we compare scheduled to earned, this is equivalent to comparing planned production to actual production. Here we see that the contractor is behind schedule overall on this work. Next, we compare EEH to actual EH. Here we see that more employee-hours are being expended than budgeted and this is an indicator that this work package is overrunning in cost from a labor point of view.

Work packaging will additionally permit direct comparison of actual quantities of materials budgeted and consumed as an additional indicator of overruns or underruns. Overall, by originally planning, estimating, and scheduling a project by work packages, the framework for total project control is established.

SUMMARY

The functions of planning, estimating, scheduling, and cost control should be part of a totally integrated cost engineering effort. If the functions are handled as separate and independent there is little possibility of ever establishing project control. Work packaging provides the ideal structure for control. Another key element of control is a well-designed code of accounts.

REVIEW

Exercises

1. Prepare a work package breakdown structure for highway construction.
2. You are a highway contractor. Develop the outline of a code of accounts you might use in this business.
3. Figure P5-3 shows data from a project's cost reports. Complete all blanks and determine overall completion for codes 03.100, 03.200 and 03.000.

Code	Description	Unit	Total est. quantity	Target EH	Placed to date	Earned EH
03.110	Forms	SF	500	5,000	500 SF	
03.120	Resteel	CWT	10	1,000	9 CWT	
03.130	Place and finish	CY	1,000	10,000	750 CY	
03.100	Slabs on grade					
03.210	Forms	SF	550	6,000	55 SF	
03.220	Resteel	CWT	10	1,000	2 CWT	
03.230	Place and finish	CY	2,500	15,000	0	
03.200	Elevated slabs					
03.000	Concrete					

Fig. P5-3

Questions In Review

1. Describe work packages associated with the erection of a structural steel frame for a building.

2. In what way does work packaging facilitate cost control during construction?
3. How are summary levels provided for in a coding system?
4. Describe types of information contained in definition, current, and historical data files.
5. What is a Work Plan?
6. Describe the output of estimating.
7. How does work packaging facilitate preparation of resource budgets?
8. What specific actions are taken by management in bid formulation?

Discussion Questions

1. Are the approaches discussed in this chapter applicable to a small specialty or general contractor?
2. Do you believe that experience in cost engineering should be a prerequisite for assignment as a construction manager?

INTRODUCTION TO CONSTRUCTION LABOR

6

INTRODUCTION

In the United States today there are thousands of contracting firms employing about 7 million workers. Of these 7 million, it is estimated that approximately half are union members and half are nonunion. The differences between union and nonunion can be translated into differences in cost. Thus, in any text on cost estimating, this subject must be addressed. It is the objective of this chapter to introduce you to union and nonunion labor as a background for the costing and productivity procedures that will be covered in Chaps. 7 and 8.

UNIONS

Background

Career construction workers are in an unusual position compared to most members of the labor force. The nature of construction work makes them transient—only occasionally can they serve a career with a single employer. Employers hire and release them as workload comes and goes so the workers necessarily keep moving to new sources of work. Thus, construction is a career field with definite uncertainty and risk. Without some form of protection, workers in this environment are subject to employer exploitation.

During the early history of the industrial revolution, the story of the American laborer was one of oppression and exploitation. Basically, all rights were retained by the employer. Various attempts by labor to organize and oppose this oppression were struck down by the police and the courts as mutinous. The United States was well into the twentieth century before the rights of the worker began to be realized. In 1932, the Norris–LaGuardia Act was passed into law. This act limited the power of federal courts to issue injunctions against union activities in labor disputes and gave workers the right to strike and picket peaceably. This was followed, in 1935, by the National Labor Relations Act (Wagner Act) which protected the right of employees to organize and to bargain collectively with their employers. It also forbade an employer from discriminating against any employee for union activity and prohibited any actions which might influence employee membership in those organizations. The act did not apply similar prohibitions against the unions. This act also created the National Labor Relations Board whose mission was to enforce the act.

The newly granted freedom of labor coupled with unsophisticated leadership within the unions soon led to strikes, restrictive practices, and criminal activity. A backlash was inevitable and by 1947, 37 states had passed some type of legislation to control labor. At the federal level, in 1947, the Taft–Hartley Act (also called Labor-Managemen. Relations Act) was passed as an amendment to the National Labor Relations Act. It included provisions to restrict unfair labor practices. The act also established the Federal Mediation and Conciliation Service and gave the President powers to act in the case of a labor dispute which imperiled national health or safety. The act still recognized the right of employees to organize and bargain collectively; its main purpose was to insure that public interest was protected in the conduct of labor affairs.

The Labor–Management Reporting and Disclosure Act was made law in 1959. This act targeted on internal union affairs to safeguard the rights of individual union members, ensure fair union elections, and prevent corruption and racketeering.

The next major legislation affecting labor unions was the Civil Rights Act of 1964 which was amended by the Equal Employment Opportunity Act of 1972. These acts were broad in scope but, for both employers and unions, made discrimination illegal.

In addition to the general legislation mentioned above, there have been a number of laws enacted which apply to any work involving federal funds. A contractor contemplating work involving federal funds is well advised to become thoroughly familiar with this legislation as well as the intricacies of federal procurement regulations because federal work is quite unlike that for private enterprise and the differences are all translatable into higher costs.

Union Organization

The International Union. An international labor organization is an association of local unions within the United States and Canada. Normally, there is an international organization for each craft or group of related crafts. The international unions affiliated with the Building and Construction Trades Department of the AFL-CIO are

1. International Union of Bricklayers and Allied Craftsmen
2. International Brotherhood of Painters and Allied Trades
3. International Association of Bridge, Structural, and Ornamental Iron Workers
4. International Association of Heat and Frost Insulators and Asbestos Workers
5. Tile, Marble, Terrazzo Finishers and Shop Men International Union
6. International Brotherhood of Boiler Makers, Iron Ship Builders, Blacksmiths, and Forgers
7. International Brotherhood of Electrical Workers
8. International Union of Elevator Constructors
9. International Union of Operating Engineers
10. Laborers International Union of North America
11. Operative Plasterers and Cement Masons' International Association of the United States and Canada
12. Sheet Metal Workers' International Association
13. United Association of Journeymen and Apprentices of the Plumbing and Pipe Fitting Industry of the United States and Canada
14. United Brotherhood of Carpenters and Joiners of America
15. United Union of Roofers, Waterproofers, and Allied Workers

The name of a union is not always totally descriptive of member crafts. For example, the carpenters union includes piledrivers, millwrights, floor layers, lathers, and drywall hangers as well as carpenters. The Teamsters Union, which is not listed as a building and trades union, is nevertheless important to construction since truck drivers are under their umbrella.

The international union maintains a full-time staff of professionals in union–management affairs to assist the local unions in their organizing work and collective bargaining. An international union may enter into a union–management agreement with a large contractor but this agreement usually will be general in nature, in effect stating that the contractor will hire union labor, use union subcontractors, and meet all local union requirements. The international union agrees, in return, to assist the employer in recruiting the labor force and helping with any disputes that may arise with local unions.

Because of the great variation in local labor agreements there is some movement toward giving the international union more authority over locals and to use national agreements which are more definitive so as to give more stability to the labor force of a contractor. At the present time, a large contractor working in many locations is dealing with countless separate local unions, sometimes having to renegotiate agreements with each annually. Since each contract renegotiation carries with it the threat of work stoppage, the impact of this system on construction progress and costs is obvious. The use of a Project Labor Agreement has evolved as a means to overcome this problem. It is a written agreement between a contractor(s) and a union(s) which establishes protective conditions for both union and contractor during the life of a project. For the contractor, a major protection is a guarantee of no strikes, work stoppages, or slowdowns.

Since a local union finds considerable advantage in being affiliated with an international union (pension funds, strike funds, professional assistance in bargaining, etc.) the primary leverage an international union has over its locals is the threat of cancelling a local's charter. This leverage is significant and the international unions have been of service to management in many instances when the locals become completely unreasonable.

Local Unions. The local union is chartered by the international union. This charter defines the geographic area and work jurisdiction for the local union. Despite their affiliation with the international union, the local unions can operate independently of the international union in the conduct of their affairs including the negotiation of collective bargaining agreements. Officers of the local union are elected and, with the exception of a business agent (BA), are not normally on a full-time salary. The business agent, on the other hand, is normally full-time; he or she is the troubleshooter, negotiator, and organizer. The BA will represent the local union with the international. The BA and the professionals from the international are charged with enforcement of the union agreements.

On each job the union will be represented by job stewards who are either appointed or elected from within the bargaining unit. Job stewards are selected from among the union employees. Technically, stewards are to perform their union duties in addition to those for which employed; practically, they spend as much as 100% of their paid time on union business such as checking union cards, investigating grievances, and protecting jurisdictional rights. Job stewards report to the business agent.

Union Training Programs

The union has two skill levels in its program: *apprentice* and *journeyman*. An apprentice is an on-the-job trainee who may also receive formal school instruction. Not all training need be done under union jurisdiction. Experience gained on nonunion work, with the armed forces, in vocational schools, or other programs may be used to meet apprenticeship requirements. The apprentice program is a structured progression of a worker through various skill levels on each of the job tasks associated with the craft. Throughout this training, the apprentice's progress is monitored, tested, and recorded. In the event the apprentice moves to another area, his or her records can be transferred.

Administration of the apprenticeship program is accomplished by joint committees with membership from both union and management. They will screen and test applicants and monitor apprentice progress. Management may sponsor apprenticeship training by providing funds, tools, facilities, and payment for the apprentices. As the apprentice moves successfully through the various levels of the training program, his or her wage rate is increased until such time as he or she is qualified for the level of journeyman. The journeyman is considered fully qualified in the trade. Journeymen may receive additional advanced training during their careers but such programs are not structured as they are for apprentice training. Certain crafts have the equivalent of a third level, that being licensing. These are crafts, such as plumbers and electricians, whose work is associated with public health or safety. There is no standardization of requirements for licensing so this procedure is not in itself an indication of skill. However, most licensing boards do require completion of apprenticeship or 4 years of equivalent experience plus a practical test.

The US Department of Labor's Bureau of Apprenticeship and Training has the responsibility of encouraging trade apprenticeship programs in the United States and assists construction management and labor in their training programs by providing standards and guidance.

Hiring Halls

The term *hiring hall* refers to the union system for referral of craftsmen (including foremen) to employers for hire. The hiring hall is a service for employers since it frees them from the problem of skilled labor recruitment. It is now common for union–management agreements to contain *exclusive hiring hall* provisions which mean that all workers must come through that referral system. In meeting its referral obligations, the union will first exhaust all local sources of qualified union labor; if this does not meet the requirement, it will get them otherwise. They may bring in union members from elsewhere in the country (travelers), or they may issue permits to transients or local citizens to work as apprentices or journeymen so that the union can meet its obligation to staff the job.

Jurisdiction

Jurisdiction defines the limits within which a given craft operates; each local is chartered for work within a certain jurisdiction. While most work activities are clearly categorized within a jurisdiction, some fall into the gray zone and lead to jurisdictional disputes. For example a local union chartered for linemen might have its jurisdiction defined essentially as all outside electrical work. Another local union in the same area may be chartered for electricians and their jurisdiction defined as all inside electrical work. When both are working on a project, problems arise at the boundary points where outside work ties into inside work. As another example, if a reinforced concrete containment vessel is to have a permanent sheet-metal lining which is used as the inside form for the concrete, a jurisdictional dispute can be raised. The steelworkers will claim the lining portion of the work as theirs because the liner is equivalent to a steel tank while the carpenters will claim the lining as theirs because it is a form. Unfortunately, jurisdictional disputes are between the unions so the contractor is helplessly in the middle. Such disputes may be settled by intervention of the international unions. However, a common form of settlement is to put a full crew of both crafts on the job.

Closed or Union Shop

The terms *closed shop* and *union shop* are used synonymously by most individuals but there is a technical difference. A closed shop exists when an employer will not hire anyone unless the person is a member of the union at the time he or she is employed. A union shop exists when an employer will employ either union or nonunion but the nonunion worker must join the union within a certain period of time. Closed-shop operations are illegal under the National Labor Relations Act while the union shop is not. In either case, the ultimate effect is the same—all workers will be union who are covered by the agreement. In areas of strong union activity, the closed shop often effectively exists, legal or not.

Agency Shop

An agency shop exists when the workforce is a mix of union personnel and nonunion personnel in the same bargaining unit and the nonunion personnel are required to pay the same fees and dues as union personnel. The rationale is that the union represents the entire bargaining unit in wage negotiations and other matters and that both union and nonunion members benefit equally. Therefore, the nonunion member should carry an equal share of the financing. The agency shop is illegal in many states.

Exclusive Recognition

With most industries, there is competition among unions to represent various employee groups. Employers cannot prevent legitimate organizing efforts of these unions but they do not have to deal with any union until it has been accepted by the affected employees (bargaining unit) as their representative. This acceptance is handled through elections supervised by the National Labor Relations Board. Once a union has gained this acceptance by majority vote, the employer is obligated to deal with it as the representative of the employees in that bargaining unit. This is *exclusive recognition*. Within a firm, there can be as many bargaining units as there are crafts or categories of employees.

Granting of exclusive recognition does not imply creation of a union shop or even an agency shop. However, it does bring the employer to the bargaining table for purposes of negotiating an agreement covering that bargaining unit and may eventually lead to a union shop.

The construction industry provides an exception to the normal organizing process because the workforce is seldom stabilized long enough to permit organization by the unions in place. Instead, the law permits a contractor to enter into a union contract prior to the hiring of any workers; in effect, the contractor exclusively recognizes a union from the beginning and commits the company to a union shop operation.

Right-to-Work Laws

Section 14(b) of the Taft–Hartley Act gives states the authority to enact laws which make it illegal for employers to make union membership a condition of employment. Approximately 40% of the states have such laws. The list includes almost all Southern and many Midwestern states. Many of these state laws also forbid agency shops.

Work Rules

In addition to agreeing to provisions concerning wages, working conditions, travel, and numerous other subjects, a union contractor must accept certain *work rules* that apply to the craft covered by the agreement. These work rules define tasks in terms of how they will be done or at what rate, or specify privileges. They invariably increase costs per unit of work but are defended by unions as being a means to protect jobs and pay for their membership. Various examples of work rules are

1. *Restriction of Production.* Example: No more than 300 bricks may be placed in an 8-hour shift.
2. *Featherbedding (Unneeded Workers).* Example: A journeyman electrician must monitor automatic equipment.
3. *Skill Inflation.* Example: Hoists (elevators) must be operated by a journeyman operating engineer when they may require no operator at all or no more than a laborer.

4. *Restrictions on Work Saving Devices.* Example: Paint sprayers may not be used; all painting will be done by brush.
5. *Restriction on Prefabricated Materials.* Example: All pipe must be fabricated on site.
6. *Set Crew Composition.* Example: One foreman and five journeymen must be used on any rigging operation regardless of size of lift.
7. *Guaranteed Overtime.* Example: Each worker will be guaranteed 6 hours overtime per pay period.
8. *Nonproductive Time.* Example: Time for changing clothes, coffee breaks, and cleanup time must be included in paid time.
9. *Unnecessary Work.* Example: All plaster must be three coat.
10. *Special Pay.* Example: Workers who show up but cannot be used because of weather or other problems will be paid for 2 hours of work.

Negotiating Agreements

Agreements are generally negotiated between the local union and the contractor. In areas of heavy union concentration, contractors have created local or regional associations and these associations may conduct negotiations for their members as a group. A typical small contractor is seldom qualified to meet organized labor on an equal footing at the bargaining table so the use of associations that are large enough to maintain a professional labor relations staff is definitely advantageous. Contractors may affiliate with one of numerous trade associations, such as the Associated General Contractors of America (AGC) or National Constructors Association (NCA) and gain further advantages from the coordination of effort that such affiliation permits. The efforts of these trade associations in the labor sector are now being coordinated by the National Construction Employers Council (NCEC).

OPEN SHOP

Open Shop, as the opposite of union shop, is the term applied to a policy of open employment. The term *merit shop* is also used. An open shop contractor will employ any individual considered qualified to perform the job whether the person is union or nonunion. In either case, the employer will not negotiate any wages or conditions of employment with the union; the company will establish its own system of wages and benefits.

An open-shop contractor may subcontract with union contractors but may insist that subcontractors have a written agreement with their unions guaranteeing that there will be no strikes, slowdowns, picketing, secondary boycotts, or work stoppages while on the project.

There always have been open-shop contractors, particularly among the small contractors and in those states with right-to-work laws. However, the movement has been growing steadily—in 1969, open-shop work accounted for about 20% of industry dollar volume; in 1979 it was closer to 60%. The national association representing open-shop contractors is the Associated Builders and Contractors (ABC). While most open-shop contractors are in the states with right-to-work laws, they are appearing in increasing numbers in the traditional strongholds of the unions.

The appeal of the open shop contractor is that work can be completed at lower cost and without schedule disruptions due to adverse union activity. However, many clients still prefer union contractors because they feel the craftsmen possess greater craft skills. In some cases, a client may prefer to go open shop but continues with union contractors because the client's operations are union and he or she fears the implications of going open shop with construction.

A number of traditional union contractors, feeling the financial pinch of open shop competition, have established *double-breasted* operations by operating a separate open-shop subsidiary. The union portion of the business operates in the traditional union areas while the open shop operates in the right-to-work states. The business becomes two separate corporations within a corporation in order to avoid violation of any laws or agreements.

Hiring

The open-shop contractor must handle all matters of recruitment alone or through employment agencies. However, both AGC and ABC now offer personnel referral services in a number of major cities. Contractors can hire either union or nonunion personnel but both categories are subject to the contractor's personnel policies, not those of a union. The contractor will establish job classifications as he or she sees fit and which may or may not bear any relationship to traditional union job classifications. The open-shop contractor may establish several levels of skill within those job classifications selected and adopt pay scales that correspond with these levels of skill. The contractor also will develop a package of fringe benefits.

It is becoming more common among large contractors for the effective value of the pay and benefits package given to the open-shop worker to be comparable with that of the union worker. The basic wage paid by the open-shop contractor may be less than that paid by a union employer but this is offset by the fact that the employee does not have union dues and assessments to pay, he or she will not face loss of pay during strikes, and in general enjoys more job security because work is not restricted along jurisdictional lines. Large open-shop contractors now offer vacation, retirement, profit sharing, and other benefits that encourage employees to remain with that firm throughout their careers.

One advantage of the open-shop system in remote areas is that the contractor can recruit workers from the area of the project who are not union members and thus help the local economies and promote better community relations. A union contractor will usually have to import much of the labor force from distant population centers since that is where union workers are.

Training

For the small open-shop contractor, worker training is usually a matter of on-the-job with no structured program for progression through established steps so skills are never assured. The larger contractors now have very comprehensive training programs for their workers which combine formal training with on-the-job training.

Cross-training of workers is unheard of in the union system while it is commonplace with the open-shop employer. Welders may be trained as electricians, carpenters as cement finishers, etc. The advantages of this system are seen in the stability of the work force. Fewer individual employees are required on a project and those employed enjoy greater continuity of work than would be the case if jurisdictional lines were followed strictly. A side benefit of this system is greater employee satisfaction and higher morale since the apprehensions associated with job-hopping are greatly reduced.

SUMMARY

The organized labor movement in the United States was necessary and inevitable. It has given dignity and saving pay to the working person. It has also given the worker new power through labor unions, a power that often has been abused, and which has taken away from management some of their rights and abilities to manage a project. The building trades unions are accused of demanding inflationary wage increases, of being discriminatory in their hiring practices, of promoting inefficiency and low productivity, and of being unnecessarily restrictive in allowing men and women into their apprentice programs. Backlash has taken place in the form of the open-shop contracting movement.

For the construction cost estimator, the difference between union and open shop can be one of dollars. In the first place, average productivity for a union worker tends to be lower than that of an equally qualified nonunion worker because of work rules. The problem is compounded by lost time and increased costs occasioned by labor disputes.

Nonworking job stewards and paid break periods also add to the cost difference. Worst case estimates have placed union labor costs 50% higher than open shop; in other areas they are considered almost equal, primarily because of skill problems that may plague the open-shop contractor.

REVIEW

Questions in Review

1. How does the Taft–Hartley Act relate to the National Labor Relations Act?
2. What is collective bargaining?
3. Relate the functions of the international union and the local union.
4. Distinguish among the terms: closed shop, union shop, agency shop, open shop, merit shop.
5. How does the union apprentice training program work?
6. How does a union contractor obtain the workers needed?
7. If an employer has recognized a union as the representative of one group of his or her employees, does this make the contractor union shop?
8. What is a right-to-work law?
9. Describe several work rules that may be part of a union–management agreement.
10. What is a traveler? A permit holder?
11. Explain double-breasted operation.
12. How will open-shop training programs under a large contractor differ from the union training programs?
13. How does the union/nonunion difference affect cost estimating?

Discussion Questions

1. Do you feel there is still a need for unions in the construction industry?
2. If you were in charge of the construction of the structures for the next Olympics complex, and it was to be a union project, what might you do to avoid the union portion of the problems encountered at the Montreal Olympics?
3. Should the federal government intervene more positively in union–management agreements?

PERSONNEL COSTS

7

INTRODUCTION

The personnel component of cost of construction has risen dramatically in recent years as demands for both higher base pay and increased fringe benefits have been met by employers. Faced with these high personnel costs employers must optimize both the size and efficiency of their work forces. They also must insure that the full costs of personnel are considered in the estimating process.

Employees can be categorized in several ways: union or nonunion, salaried or hourly wage, direct cost or overhead, and professional or nonprofessional. Each of these categorizations may enter into the planning and estimating process, but the categories of salaried and hourly wage are most significant from the estimator's standpoint.

A salaried employee is one with a fixed periodic salary, usually expressed as a monthly or annual salary. Most of the personnel in the home office, both professional and nonprofessional, will be salaried. In addition, at the job site, the noncraft and supervisory personnel usually will be salaried. Hourly wage employees are those paid per hour of work so their periodic paycheck will be directly related to the number of hours worked during a pay period. On construction projects, the construction workers, both skilled and common labor, will normally be hourly wage.

For the salaried employees, productivity is important but does not play a part in the estimating process. Staffs of salaried individuals are established based upon functions which must be performed. Productivity considerations enter into the staffing decision but, once the staff is established, the estimator is concerned only with the number and periodic cost of the staff members.

For the hourly wage employee, productivity is vitally important to the estimator. Here, the estimator identifies a quantity of work to be performed. He or she then determines the employee-hours to accomplish the task, a determina'ion which is possible only if the labor productivity is known. Once the productivity and employee-hours are established, the estimator can then determine the production cost of labor by multiplying employee-hours by cost per hour. In summary, for hourly wage employees, the labor production component of cost is found as

$$\text{Labor cost} = \frac{\text{(Quantity of work)}}{\text{(Productivity)}} \times \text{(Cost per unit of time)}$$

With some exceptions, the hourly wage employee costs will become part of the direct costs of the project. The salaried personnel costs generally will be either a part of job overhead (indirects) or part of general overhead (main office or distributed) costs depending upon the employee's function and location.

The true cost of an employee to an employer is a combination of the employee's basic wage or salary, mandatory taxes and insurance, plus the cost of any other benefits. Each of these will be discussed in turn.

BASE WAGES AND SALARIES

The base figures to be used for salaried personnel in an estimate can be obtained from the personnel department of the firm. Hourly wage rates to be used for labor will be union scale if the project is to be under union contract or prevailing nonunion scale if the project is open shop. On federal projects, contractors are required to pay prevailing wages in the area under the provisions of the Davis–Bacon Act. This is generally interpreted to mean union scale whether the project is union or open shop. In any event, wage rates are area sensitive so the estimator must insure that valid data from the area of the project being estimated is used. This data can be obtained from local employment offices, unions, the US Department of Labor, state labor agencies, potential subcontractors in the area of the project, or from commercial cost services.

Current salary or wage data will be used only if the project being estimated will be completed during the period that these figures are valid. For longer projects, the estimator must project what the salary and wage figures will be. One common approach is to project these costs to the midpoint (in time) of construction. This is approximately the point at which staffing will be a maximum on the project and the estimator, in using this approach, assumes that those figures are an average of wage and salary scales that will prevail over the life of the project. On projects of extremely long duration, such as nuclear power plants or large dams, the estimator may choose to project wage and salary scales for each of several phases of the project. For Phase I, the estimator may project rates 2 years forward; for Phase II, 3 years; etc. In making these projections the estimator must rely upon historical data on salaries and wages and any information available from governmental, union, business, or other cost sources. The estimator must be extremely cautious in the estimation of costs of a union project if union contracts will be renegotiated between the time of the bid and the completion of the project. If the union demands during renegotiation significantly exceed the figures assumed in estimating a project the contractor can be hurt in two ways. First, if the union demands are accepted, the contractor will surely lose money on the labor component of cost. Second, if the contractor does not accept the union demands and lengthy negotiations result, the project will be delayed, again adding to costs since certain project costs (overhead items, interest, etc.) continue throughout the delay period. The contractor may be able to eliminate these potential problems by negotiation of a project agreement for the project in which wage scales and fringe benefits are defined and fixed (or adjusted according to a pre-agreed formula) during the course of the project.

During the 1970s wage rates increased from 6% to 10% a year, depending upon the location and the year. Opinions for future escalation tend to vary over the same general range.

PAYROLL BURDEN ITEMS

A common synonym for employer-paid taxes, insurance, and other fringe benefits is *payroll burden* since the cost of these represents an added burden to the employer that can range from 10 to 35% or more of base wage cost. Some benefits are shared in cost by employee and employer. In such instances, only the portion borne by the employer is included as payroll burden since the employee must pay his or her share from his or her base wages.

There are two basic approaches for accounting for payroll burden during the estimating process. Under the first method, the costs of these items are added to the base salary or wage figures. That is, for salaried employees, the payroll burden items are converted into an annual cost and added to the annual salaries. For the hourly wage employees, they are converted into an equivalent hourly cost and added to the base hourly cost. Under the second method, payroll burden items are lumped together and treated as a main office or job indirect expense. A combination of the two is often used by a construction estimator. The first method will be used for those payroll burden items which vary with hours worked and which are keyed to the base wage, such as social security, unemployment insurance, and worker's compensation insurance. The cost of other items are combined and handled as a job indirect or main office distributed cost. This approach is adopted so that the true incremental labor cost is available should any changes be introduced later in the work package.

Following discussion presents a description of each of the major payroll burden items that will be encountered by a construction contractor.

Social Security

Social security is a federal program designed to bring eventual retirement, medical, survivor, and other benefits to employees or their beneficiaries. The program is funded jointly by employees and employers, each contributing an equal amount each pay period. Social security coverage is mandatory for employees of most businesses, including construction.

Social security premium rates are established by law and are subject to annual adjustment. The rates are a combination of a percentage multiplier and an income ceiling. The percentage multiplier is applied to all income up to the ceiling, with no premium on amounts over the ceiling. For 1980 the premium rate was 6.13% on all income through $25,900. Assuming that an employee made $30,000 in 1980, the annual premium was

$$(0.0613) \ (\$25,900) \ = \ \$1587.67$$

Note: All amounts over $25,900 are not taxed. This amount would be paid by *both* the employer and the employee so that the total premium received by the federal government is $3175.34 for that employee. For an employee who received only $12,000 in 1980, the annual premium the employee paid was

$$(0.0613) \ (\$12,000) \ = \ \$735.60$$

All payments made are credited to the employee's individual social security records in Washington. The estimator is concerned only with that part of the social security cost borne by the employer since that paid by the employee is deducted from base salary or wages.

For long term projects an estimator must make a projection of future social security costs. Table 7-1 provides a schedule of contributions as contained in 1977 amendments to social security legislation. Increases in ceilings for 1982 and beyond may be estimated based upon assumed annual increases in wages. For example, if one assumes an annual increase in average wages of 6% after 1981, the ceiling for 1989 (8 years later) may be estimated as

$$(\$29,700) \ (1.06)^8 \ = \ \$47,337$$

Unemployment Insurance

Unemployment insurance is the second type of insurance coverage that most contractors find mandatory by law. As its name implies, it is

TABLE 7-1. Social Security Rates

Year	Multiplier	Ceiling
1978	6.05%	$17,700
1979	6.13%	$22,900
1980	6.13%	$25,900
1981	6.65%	$29,700
1982	6.70%	$31,800*
1983	6.70%	*
1984	6.70%	*
1985	7.05%	*
1986–89	7.15%	*
1990–2010	7.65%	*

*Estimated. The ceiling for this and following years will be adjusted to account for increases in average earnings and benefits.

a program designed to provide protection to workers during times of unemployment. The unemployment insurance program has its basis in both federal and state laws. In operation, the program is administered at state level although backup funds for the state programs are maintained by the US Treasury in accounts reserved for the individual states.

The employer makes the only premium payments for unemployment compensation except in Alabama, Alaska, and New Jersey where the employees are subject to a portion of the tax. There are two components to the premium. The first component is for payment of federal costs in connection with the program and for maintenance of a federal level loan fund used to back up the state funds in time of high regional unemployment. This component is 0.7% of the first $6000 of wages. The second component of the premium goes into the state's account from which all benefits are withdrawn. This percentage is variable but will average no lower than 2.7% of the first $6000 of wages. The percentage multiplier is an average considering all employers in the state. For a given employer it can range from zero to more than 7% since employers are *experience rated*. In the long run employer contributions must equal payments to former employees who have filed for and received unemployment benefits following release from the firm for lack of work. Thus, those with low layoff records pay less than those with high records. Rates also vary by states because benefits vary by states. A state can increase both the average rate and the ceiling if necessary to maintain adequate fund reserves. Rates are announced annually so an estimator must be certain that the latest information for all states within which the company works is on hand. It is difficult to project future contributions because they are dependent upon employer layoff experience. An estimator may logically assume that unemployment benefits will tend to increase with the overall cost of living and project future costs on the basis of present costs increased by expected inflation.

The nature of unemployment insurance contributions leads to a note of caution. It sometimes happens that an employer intending to fire an employee for cause will instead simply agree to release the employee as if work were no longer available so that the former employee will not lose unemployment benefits. The employer must realize that the firm will eventually pay for those benefits and thus add another overhead cost to the operation.

Worker's Compensation Insurance

This insurance is designed to provide protection to employees who are killed, injured, or suffer health problems due to job-related accidents or conditions. Each state has its own law covering worker's compensation (WC) insurance. Unfortunately, these laws, although similar in principle, are quite different in detail so that costs associated with this insurance program must be related to specific states. As a general rule, contractors are required to provide this insurance for their employees although very small contractors may find themselves exempt in certain states. As a matter of good practice, all contractors should provide this insurance because failure to do so exposes them to possible legal action by employees injured on their jobs. A feature of worker's compensation laws is that an employer, in carrying this insurance, is assuming liability without fault. The contractor assumes the liability by purchasing the insurance for the employee. In return, an employee injured on the job cannot bring further suit against the employer for damages, although he or she can sue a fellow employee, another contractor on site, or other individual or business other than the employer that may have been involved.

Depending on the state, an employer may obtain WC insurance from a commercial insurance company, a state operated insurance fund, or the company may qualify for self-insurance. Whatever the source of coverage, benefits included must be those established by the state law.

The rates paid by each contractor are a function of the state involved, the craft being insured, the source of insurance coverage, and the accident experience of the contractor. The starting point for the rate structure is a series of tables prepared by the National Council on Compensation Insurance. These tables are revised annually. The tables are advisory only so states and insurance companies will modify them to reflect their respective situations. Once modified, they become the base rate structure for workers being insured in that state. This base rate structure normally will apply directly to all small contractors and to new contractors, regardless of size, since they have not had enough accident exposure for experience rating. For the remaining contractors the rates listed in the table may be adjusted based upon past experience.

The adjustment process will result in a multiplier to be applied to the base rates of the insurer. This multiplier will be greater than 1.0 for those firms with poor accident records and will be less than 1.0 for the firms with good records. In Texas, for example, the contractor is evaluated on experience during a continuous 3-year period, this period beginning 4 years before the year for which the rate is being established and continuing through the second year prior to the rate year. In other words, a contractor's rates for 1980 will be based upon the company's experience in 1976, 1977, and 1978. The experience for the first prior year (1979 in the example) is not used since the rate application is prepared during that year and data for the year will be incomplete.

The cost implications of poor safety records cannot be overemphasized. For example, the worker's compensation base rate for structural steel workers in numerous states exceeds $30 per $100 of payroll. A contractor in a state with a $30 base rate who has a good safety record may earn a multiplier of only 0.5 and pay only $15 per $100 while a contractor with a poor record and a possible multiplier of 2.0 will pay $60 per $100.

All rates are expressed in terms of dollars per hundred dollars of payroll which is equivalent to expressing it as a percentage of payroll. Thus, if a contractor has an annual carpenter payroll of $100,000 and the insurance rate for carpenters is $1.12/$100, the annual insurance premium is $1120.00 for that particular craft. Payroll includes base wages plus any other compensation considered taxable as wages. Overtime pay is not taxed fully; it is reduced to straight time pay for hours worked. WC rates can be expected to increase at about the same rate as the overall cost of living. In 1978, the average construction contractor paid about 6.5% of payroll into worker's compensation.

A contractor operating in more than one state will be subject to more than one rate structure so estimators must insure that rates being used are for the proper state.

Table 7-2 is an extract from a table of average WC rates applicable to construction trades and effective July 1, 1979.

Vacations

For salaried employees, an employer will normally allow a certain number of days paid vacation each year. Very often, the amount allowed will be based upon years of service with the company. Since the salaries of salaried employees are normally handled in estimates by using an annual amount, the cost of any paid vacations are automatically included and the estimator does not need to be concerned with a separate accounting for vacations. This is not the situation with hourly wage employees since productivity calculations always assume that the worker or crew is physically on the job. Therefore, any paid vacation provided by an employer for hourly wage employees is an added cost which must

TABLE 7-2. Workmen's Compensation Rates

RATE IS PER $100 PAYROLL. COMPILED BY HERBERT L. JAMISON CO. INSURANCE ADVISERS AND AUDITORS, NEW YORK, N.Y.

Effective July 1, 1979 CLASSIFICATION OF WORK	Ala.	Ak.	Ariz.	Ark.	Cal.	Colo.	Conn.	Del.	D.C.	Fla.	Ga.	Hi.	Idaho
Carpentry—1, 2 family residence	3.76	7.09	10.57	5.54	8.43	5.37	10.09	6.11	14.41	10.22	4.77	12.22	6.86
Carpentry—3 stories or less	4.05	8.62	11.09	5.65	8.43	6.48	10.09	6.11	14.41	10.22	4.77	12.44	6.86
Carpentry—interior cab wk.	2.58	6.19	10.04	3.60	8.43	3.78	6.40	6.11	10.99	7.47	3.34	6.41	4.25
Carpentry—general	4.28	8.90	18.90	6.23	10.47	5.59	12.50	6.11	18.37	14.53	7.22	21.39	10.15
Chimney—construction brick con	8.13	24.52	22.24	20.05	17.02	11.24	44.17	A	78.09	35.25	10.87	29.45	16.25
Concrete work—bridges culverts-C	5.87	11.08	16.20	9.99	17.02	8.33	16.51	8.90	25.50	18.29	6.21	12.85	7.93
Concrete work—dwelling 1-2 family	2.51	8.26	8.51	3.10	5.17	3.78	8.91	8.90	23.69	7.39	2.45	10.16	4.20
Concrete work—N.O.C.	4.85	9.14	13.80	5.08	8.56	6.86	8.52	8.90	24.93	14.42	5.12	13.80	6.93
Concrete or cement work—floors, sidewalks	3.68	6.70	8.25	2.79	4.12	3.89	7.67	8.90	12.94	10.02	3.38	8.40	3.60
Electrical wiring—inside	2.29	5.41	7.22	3.00	4.28	3.18	3.59	2.18	10.67	6.03	2.73	7.86	3.42
Excavation earth N.O.C.	5.06	7.89	8.15	5.54	6.14	4.66	8.27	9.30	17.01	10.58	6.01	19.01	10.06
Excavation—rock	4.95	8.56	15.03	6.18	10.34	6.42	11.08	9.30	48.25	15.86	8.57	13.16	6.91
Glaziers	5.09	8.74	10.56	6.54	8.20	5.83	9.87	6.33	20.81	9.88	7.03	12.34	6.85
Insulation work	4.62	8.07	11.50	4.53	7.52	5.87	6.81	6.11	13.86	10.64	4.67	6.89	8.13
Lathing	2.28	5.46	7.09	3.32	5.28	3.87	5.82	5.88	11.12	7.56	2.69	6.04	3.88
Masonry	3.18	10.44	11.71	3.50	8.75	5.57	11.01	7.78	23.95	9.41	3.96	10.16	5.52
Painting and decorating	3.36	8.09	7.25	4.49	7.98	4.33	9.73	11.20	14.85	8.10	3.91	8.25	6.75
Pile driving	11.83	24.50	34.72	18.81	19.35	15.68	8.42	12.71	67.37	26.44	3.91	35.57	17.05
Plastering	3.74	7.78	18.83	4.22	9.27	5.46	6.84	5.88	12.82	9.42	4.67	8.85	5.93
Plumbing	3.10	6.90	7.35	3.34	5.17	3.51	4.24	4.09	12.05	6.70	3.71	7.77	3.48
Roofing	7.88	20.00	24.58	11.41	16.42	15.45	18.13	10.30	33.80	22.08	8.30	27.29	15.38
Sheet metal work—erection—inst. & repair	3.20	6.28	9.87	3.70	5.62	4.28	7.71	10.30	13.11	7.46	5.04	5.24	4.13
Steel erection—doors and sash	3.37	8.69	9.93	4.34	6.32	4.39	6.06	12.26	21.50	6.43	3.91	6.82	5.56
Steel erection—interior ornament	3.37	8.69	9.93	4.34	6.32	4.39	6.06	12.26	21.50	6.43	4.67	6.82	5.56
Steel erection—structure	6.53	36.15	42.96	16.34	17.93	15.73	27.98	12.26	59.37	30.10	10.73	37.87	16.08
Steel erection—dwelling 2 stories	10.23	25.79	38.80	10.92	17.93	11.99	19.17	12.26	59.19	20.29	12.98	32.65	18.58
Steel erection—N.O.C.	9.85	22.18	21.07	12.05	16.23	15.79	26.67	12.26	73.47	21.19	12.39	32.47	4.13
Tile work—interior	2.63	2.99	8.14	3.20	4.97	3.32	4.61	4.09	32.94	5.94	2.53	6.60	4.47
Timekeepers and watchmen	3.37	8.02	7.39	4.58	6.99	4.46	7.72	A	21.56	9.64	4.68	11.36	5.30
Waterproofing (brush) interior	3.36	8.09	7.25	4.49	7.98	4.33	9.73	A	14.86	8.10	3.91	8.25	6.75
Waterproofing (trowel) interior	3.74	7.78	18.83	4.22	9.27	5.46	6.84	A	12.82	9.42	4.67	8.85	5.93
Waterproofing (trowel) exterior	3.18	10.44	11.71	3.50	8.75	5.57	11.01	A	23.95	9.41	3.96	10.16	5.52
Waterproofing (pressure gun)	4.85	9.14	13.80	5.08	8.56	6.86	8.52	A	24.93	14.42	5.12	13.80	6.93
Wrecking	15.63	30.72	65.10	20.75	A	24.36	54.19	25.42	51.55	52.29	19.11	51.79	36.68

(Reprinted from *Engineering News-Record*, September 20, 1979, copyright McGraw-Hill, Inc., all rights reserved.)

be accounted for in addition to the base wages. Vacations for union employees are normally provided through payment to a union vacation fund. In such cases, the amount of these payments becomes the vacation cost. For the nonunion employer who provides paid vacations for hourly wage employees, the added cost equals the number of days provided multiplied by the daily wage rate. Vacation pay is subject to social security, unemployment, and worker's compensation tax, so the premiums for those items must be added unless the maximum withdrawals have already been satisfied.

Medical Insurance

It is quite common for employers to make group medical insurance plans available to their employees. An employer may pay the full cost of medical insurance for the employees or just a portion. Whatever the amount contributed by the employer, it is an added cost that must be accounted for. Since the contribution may vary from employee to employee, an estimator would normally use an average contribution per employee in estimates since estimates do not identify individual workers. For union contractors, contributions to a union health and welfare fund would replace contributions to a medical insurance program for included union employees.

Other Insurance

Like medical insurance programs, group life and/or accident insurance programs may be available to the employees. Any employer contributions to such programs or administrative costs in connection with them would be handled in a manner similar to medical insurance.

Other Benefits

Among other possible benefits funded by an employer are pension plans, on-duty training programs, off-duty educational assistance, sick leave, etc. For the union contractor, these will take the form of contributions to respective union funds.

For estimating purposes, the costs of these additional payroll burden items would be accounted for by one of the methods discussed for other payroll burden items.

SPECIAL PERSONNEL COSTS

There are several additional personnel costs which do not fit into the categories of base wages or payroll burden. These costs are discussed below.

Transportation Costs

For jobs in overseas areas, an employer will normally pay the cost of commercial transportation of the employee (and perhaps family members) from their home country to the country of employment. Even in the United States, the remoteness of a construction site may dictate that an employer pay mileage to and from the construction site on a daily or weekly basis. Some union contracts will require that all union employees be paid a mileage fee between the project and the employee's home or the union hiring hall.

Travel Time

Associated with mileage payments can be a payment for time spent in travel. For example, an employee might receive 8 hours pay each day but is permitted to take 1 hour of that time in travel. The effect is to reduce the employee's working hours to 7 and to increase the effective hourly base pay by one-seventh. It is recommended that this adjustment be handled by an estimator as an increase in base pay for each of the 7 hours on the job so as not to throw off productivity considerations which are keyed to actual hours at work.

Shortened Shift Time

On multiple shift operations an employee may be paid for 8 hours of work although he or she is required on the job for only 7 or $7\frac{1}{2}$ hours on the swing or graveyard shifts. Since productive time is reduced, this condition should be treated in the same manner as travel pay by adding an amount to the base hourly pay.

Subsistence or Per Diem

An employer may find it necessary to provide either living quarters and eating accommodations for employees or to provide an additional amount of money each day (per diem payment) so that the employees can obtain accommodations in a nearby community. This type of payment is normally reserved for those situations when employees cannot return daily to their homes.

Area Premium Pay

When working conditions are unfavorable or the project is remote or overseas, an employer may pay a bonus to employees who accept work under those conditions.

Overtime and Overtime Premium Pay

An hourly wage employee who is required to work hours in excess of the normal is entitled to receive overtime pay which is usually computed at $1\frac{1}{2}$ to 2 times regular base pay for each overtime hour worked. Additionally, on some union projects, the agreement requires overtime premium pay for all work performed on weekends and holidays whether or not those are true overtime periods. For example, weather may stop a project on Thursday and Friday so the contractor instructs the workers to report on Saturday and Sunday to make up time, those being the fourth and fifth workdays of the current week. A nonunion contractor would probably not pay overtime premium for Saturday and may or may not pay it for Sunday, while a union contractor would probably pay premium for both days.

There may be scheduled or casual overtime on a project. Scheduled overtime is that which is planned as a standard feature of the project. This type of overtime is common on projects which are weather sensitive, such as paving operations. Casual overtime is occasional overtime made necessary by special conditions of the job. For example, casual overtime may be necessary to insure proper finishing of fresh concrete work before it sets.

To handle scheduled overtime, an assumption must be made as to the extent such overtime will be used. Then, the added cost of this overtime should be converted into an equivalent added cost for every hour worked. For example, assume that 4 hours of overtime work is anticipated per craftsman per week. Also assume that the hourly base wage rate is $12 per hour and overtime premium is $1\frac{1}{2}$ times base.

$$\begin{aligned}
\text{Pay for first 40 hours} &= (40 \text{ hours}) (\$12/\text{hour}) &= \$480 \\
\text{Overtime hours pay} &= (1\tfrac{1}{2}) (4 \text{ hours}) (\$12/\text{hour}) &= \underline{72} \\
\text{Total pay per week} & &= \$552 \\
\text{Average pay per hour} &= (\$552) \div (44 \text{ hours}) &= \$12.55
\end{aligned}$$

To account for casual overtime pay the estimator may choose to add a few cents to the base pay or a single lump sum can be added to the overall labor estimate to account for all anticipated casual overtime.

Paid Hours and Productive Hours

In earlier discussion it was mentioned that a worker may be paid for nonworking time—examples were given of shortened shifts and travel time. Thus, there can be a difference between paid time and productive time. Productive hours are of ultimate importance in scheduling so conversion of costs to cost per productive hour is recommended. A procedure for doing this is incorporated into the worksheet to be presented next.

MAKING CALCULATIONS

Developing a Worksheet

As with most of the estimating processes, an estimator is well-advised to develop a checklist or worksheet for handling the calculation of effective personnel costs. Figure 7-1 is a sample worksheet for hourly wage employees which organizes the calculations so that an estimator can isolate the equivalent hourly cost of those items which relate directly to hours worked from the cost of fringe benefits and other personnel costs which are more of an annual cost, not a function of hours worked. This approach allows the estimator to include the latter group of costs in job indirects or in main office distributed costs if he or she chooses or to convert them into an equivalent hourly cost. Detailed as this worksheet is, it still provides only an estimated hourly cost since the estimator must make input assumptions concerning hours worked per year and overtime in addition to projecting wages, rates, and ceilings to the proper point. However, having organized the process by means of the worksheet the estimator is more likely to remember everything and develop a reasonably accurate figure than would be the case if he or she were to arbitrarily apply a percentage factor to base wages to account for all other costs.

Determining Days/Hours Worked per Year

One of the key entry items for use of the worksheet in Fig. 7-1 is the number of working hours per year. Following is an example of calculations to determine this figure. The estimator must adjust the quantities used to fit the situation.

Days in year		365
Nonproductive days		
Weekends (52 weeks) (2 days/weekend)	104	
Holidays	11	
Vacation	10	
Sick	5	
Weather	5	
	135	
		135
Productive days/year		230
Productive hours/year = (230 days) (8 hours/day)		1840

HOURLY WAGE WORKSHEET

Craft:

Project _____

Location _____

Date for which determined _____

A. _____ Basic wage rate ($/Hr) F. _____ Worker's comp ($/$100)

B. _____ Social security rate (%) G. _____ Straight hours annually

C. _____ Social security ceiling ($) H. _____ Annual hours overtime

D. _____ Unemploy. ins. rate (%) J. _____ Overtime premium multiplier

E. _____ Unemploy. ins. ceiling ($)

K = annual pay = (G × A) + (H × A × J) = ... $ _____

L = annual pay subject to worker's comp = $ _____

M = Social security contributions = $\frac{(B \times C)^*}{100}$ or $\frac{(B \times K)^*}{100}$ = $ _____

*Use lower of two values

N = unemploy. ins. contributions = $\frac{(D \times E)^{**}}{100}$ or $\frac{(D \times K)^{**}}{100}$ = $ _____

**Use lower of two values

P = Worker's comp contributions = $\frac{(F \times L)}{100}$ = ... $ _____

Q = total costs relating to hours worked = K + M + N + P = | $ |

Other contributions (annual costs):

     Medical insurance . . . . $ _____

     Other insurance . . . . $ _____

     Paid vacation . . . . $ _____

     Travel pay . . . . $ _____

     Subsistence/per diem . . . . $ _____

     Other premium pay . . . . $ _____

     _____ $ _____

     _____ $ _____

R = Total other contributions . . . . $ _____

S = Total annual costs = Q + R = ... | $ |

T = Annual hours shift/travel time reduction = _____

U = Equivalent cost per productive hour = $\frac{Q^{***} \text{ or } S^{***}}{G + H - T}$ = $ _____

    *** Use Q if other contributions (R) handled
       as job indirect cost. Use S if all
       costs converted to hourly cost.

Fig. 7-1   Hourly wage worksheet.

Projecting Future Wage Scales, Taxes, and Insurance Rates

An estimator must take advantage of trade magazines, government publications, and cost service publications so that he or she can keep abreast of all cost trends. From these sources the estimator will learn what is being done in the industry and what is projected. In the case of wage scales, the most recent data on wage settlements will be an indicator of the periodic growth rate of the wage and fringe benefits package.

SPECIAL REQUIREMENTS FOR WORK IN OVERSEAS AREAS

Almost every nation has enacted tax and social legislation covering workers in their country. A contractor considering work in an overseas area must research these laws and develop cost information similar to that used for US workers. A US contractor operating overseas actually may be subject to three sets of requirements

1. US requirements covering workers who are US citizens
2. Host country requirements covering local national workers and, in some cases, the entire labor force
3. Third country requirements covering labor imported from a nation other than the United States or host country

SUMMARY

The total cost of maintaining a person on the payroll is the base wage or salary plus employer contributions to social security, worker's compensation, unemployment insurance, vacations, and other benefits. The cost of those added payroll burden items is significant and constantly changing so it is essential that an estimator be up-to-date on all of them. An estimator also should deal in costs per productive hour when costing field workers since labor requirements are in these terms. Productive hours can be less than paid hours if paid travel time or shortened shift time reduces hours on the job.

It is always advisable for an estimator to use checklists or worksheets in developing costs. These worksheets catalog all possibilities and help prevent serious omissions. They also may serve as coding sheets if computerized procedures are employed.

REVIEW

Exercises

1. What do you project the social security ceiling will be in 1995?

2. In 1978 workers received wages in the amounts listed below. What amounts of social security taxes were paid by the employee and the employer?
   a. $16,000
   b. $20,000

3. In 1979 a structural steel worker in Colorado earned $20,000 in wages. What amount was paid by the employer and the employee for workmen's compensation insurance? Assume rates are as given in Table 7-2.

4. The base wage of a worker is $10.00 per hour. What is the cost per productive hour under each of the following conditions?
   a. The worker will work the graveyard shift and will be paid for 8 hours but work only 7 hours.
   b. The worker will work 6 hours per week of scheduled overtime. Overtime premium rate is $1\frac{1}{2}$.

5. A worker receives a basic wage of $6.00 per hour during a normal work week of 40 hours. The worker is put on a swing shift which calls for a shortened shift time of $\frac{1}{2}$ hour. He or she also works an average of 4 hours overtime each week at a premium rate of $1\frac{1}{2}$. What is the cost per productive hour for this employee?

6. In a mountain community of Colorado the construction season for residential construction extends from May 15th to about September 30th. Assuming a normal 8-hour workday and 8 days of lost time due to weather during that period, calculate the number of productive hours available per year.

7. Calculate the total cost per productive hour of each of the listed crafts assuming the following:

Location: Arkansas Year: 1979	Concrete Worker	General Carpenter	Plumber
Basic wage rate	8.45	9.75	11.25
Working days per year	200	200	200
Travel time, hours/day	1	0	0
Shift time reduction, hours/day	0	$\frac{1}{2}$	
Average overtime, hours/week	0	2	0
Overtime premium multiplier	NA	$1\frac{1}{2}$	NA
Miscellaneous contributions, total value/year	$1200	$1400	$1600
Unemployment insurance rate/ceiling	3.4%/$6000	same	same

Use Tables 7-1 and 7-2 for social security and worker's compensation rates.

8. Develop a cost worksheet for salaried personnel similar to that described in the chapter for hourly wage personnel. However, the end cost should be cost/month.

Questions in Review

1. For a typical contractor what are the differences between hourly wage and salaried employees?
2. How does productivity enter into the costing of a salaried employee?
3. How are base wage rates established?
4. Define payroll burden.
5. When a cost is distributed, what does this mean?
6. What is the purpose of social security? Who pays for it?
7. How is the unemployment insurance program funded and administered?
8. How is the rate structure for worker's compensation established? Who pays for it?
9. Explain shortened shift time.
10. Can an employee earn overtime pay without working overtime? Explain.
11. What is the difference between casual and scheduled overtime?

Discussion Questions

1. Why should an employer accept responsibility for employee accidents? Who should be responsible?
2. Should a contractor bidding on a long-term project be expected to project future wage rates, social security rates, unemployment compensation rates, and worker's compensation rates? Can you think of any way the contractor could be relieved of this risk?
3. Would you suggest that the overall system for payment of employees be changed to give them a higher base wage and then deduct costs of all fringe benefits from that wage rather than use the present system of wages plus benefits?
4. How are job safety and construction costs related?

LABOR PRODUCTIVITY

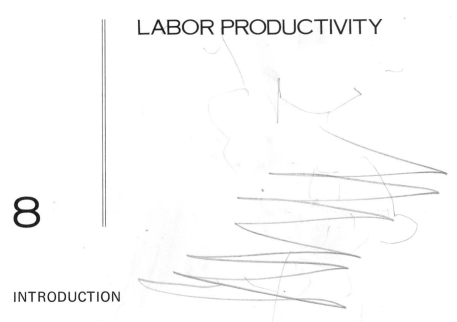

8

INTRODUCTION

You will recall from Chap. 7 that the basic equation for determining labor cost was

$$\text{Labor cost} = \frac{(\text{Quantity of work})}{(\text{Productivity})} \times (\text{Cost per unit of time})$$

Determination of the cost of a crew per hour was discussed in detail in that chapter. All that remains to complete your understanding of the labor costing process is an understanding of productivity.

Both the quantity of work to be performed and the crew cost per hour can be established with considerable accuracy. *The real variable in the equation is productivity.* The extent of productivity variation on the same type of work is well illustrated by the data in Table 8-1 which was compiled from a study of productivity on nuclear power plants.[1] These data show that the ratio between high and low productivity figures is almost 3 to 1. Such variations exist because many factors influence human productivity. Identification and evaluation of these factors is one of the greatest challenges for an estimator. Yet, it is a challenge that must be met because every error in productivity estimation causes an inverse error in the actual labor cost. It is the purpose of this chapter to identify the factors that influence human productivity and to present an approach for evaluating their effect.

Numerous studies have been conducted on labor productivity, each designed to catalog the manner in which typical workers spend

---

[1] Budwani, Ramesh N., "Important Statistics on Engineering and Construction of Nuclear Power Plants," *Transactions*, American Nuclear Society Topical Meeting, 1976, pp. I-5-1 to I-5-14.

TABLE 8-1. Variation of Labor Productivity

Work Item		Employee-Hours Expended		
		Reactor Bldg	Turbine Bldg	Auxiliary Bldg
Concrete EH/CY	High	5.50	4.10	3.70
	Low	2.10	1.50	2.00
Resteel EH/Ton	High	60.00	44.00	42.00
	Low	22.00	18.00	22.00

Ramesh N. Budwani, "Important Statistics on Engineering and Construction of Nuclear Power Plants" (Proceedings of the topical meeting, Nuclear Power Plant Construction, Licensing and Startup, September 13–17, 1976), p. 1.5–12.

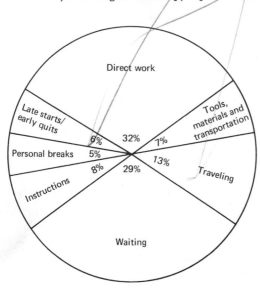

Fig. 8-1 Breakdown of a typical craftman's hour. (Courtesy Marjatta Strandell and American Association of Cost Engineers)

their time on a job and to then use this information as a basis for development of methods for improving productivity. Figure 8-1 shows the summary of one such study,[2] the data presented representing an average for a number of crafts. Looking at these figures, you may be alarmed to find that the average craftsman was productive only about one-third of the time. Averages varied from craft to craft, with some as low as 26% and others as high as 40%, but in all cases, well over half of the workday was spent in waiting, traveling, breaks, and other nonproductive activity. The data of Fig. 8-1 applies to large power plant projects but is considered representative of large industrial type construction. On smaller projects, one can normally expect greater percentages of

[2] Strandell, Marjatta, "Productivity in Power Plant Construction," *Transactions*, American Association of Cost Engineers, 1976, pp. 334–337.

productive time, perhaps averaging closer to 40%. On the other hand, productive time on "super projects" will tend to be less, as low as 15 to 20% on the average. With so little time spent in productive work it can be shown that a very small shift in time distribution can have the effect of significantly increasing or decreasing productivity. For example, using the data of Fig. 8-1, the productive time (direct work) of an average craftsman during an 8-hour day is

Productive time $= (0.32)\,(8\text{ hrs})\,(60\text{ min/hr}) = 153.6$ minutes

To increase productivity by 10% would involve the addition of only 15 to 16 minutes to the productive time each day. Thus, if a supervisor could reduce waiting time, travel time, instruction time, or other nonproductive time by that amount each day, productivity would increase by 10%. Imagine what the effect of an across-the-board increase of 10% would do to costs. In 1978, the bidding volume for major construction (not including 1 and 2 family dwellings, buildings under $500,000, and heavy and highway construction under $100,000), as reported by *Engineering News-Record*, was in excess of $55 billion. Assuming labor costs account for about 40% of project costs, $22 billion of the $55 billion was labor cost. If productivity could have been improved by just 10% the savings in that year would have exceeded $2 billion.

Productivity is extremely sensitive to the quality of management and supervision. Large contractors with a number of supervisors in their employ will admit that they experience considerable variation in the average productivity generated by their different construction superintendents. In fact, if productivity variation among managers were cataloged, an estimating group would have a basis for adjusting productivity and cost estimates based upon identification of the intended construction supervisory staff.

Figure 8-2 identifies many of the causes of low productivity. It can be seen that many of these detracting factors are of a type which good management can eliminate or ameliorate. Obviously, it never would be possible to eliminate all nonproductive time. So, what is the practical limit of direct work time that can be achieved? There is no exact answer, of course, but research has suggested that it can approach 55%.[2] If true, there is ample room for productivity improvement among most labor crews.

What are some of the actions that management can take? Time lost to workers while instructions are being given or while they are waiting are prime targets for management improvement actions. If a company develops and maintains a manual of construction procedures that are standard throughout the firm, the necessity for new and detailed instructions for each work item can be greatly reduced. Such standard

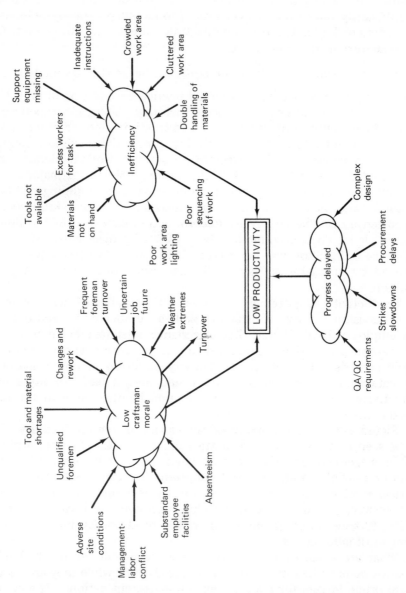

Fig. 8-2 The low productivity problem.

operating procedures would outline the actual methods for accomplishing all common tasks whether it be warehousing of materials or the pouring of concrete in a slab. Waiting time can be reduced by better scheduling of work to insure that the labor, materials, and equipment are available simultaneously and one element is not waiting on the other. Efficient site layout is extremely important. The area for check-in and check-out (brass alley) of employees should be located to minimize travel time to and from work locations. Laydown areas for construction materials should be as close as possible to place of use to prevent costly movement time or double handling. Construction water, air, gas, and other service lines should be located for quick and easy access. These are but a few of the many management actions possible, all designed to convert nonproductive time into direct work time.

On some projects, the use of scale models can be employed to determine construction sequences and procedures in advance of construction to maximize direct work time and reduce time lost by delays or in correcting mistakes caused by poor prior planning. The structure is built step-by-step in model form to insure that equipment and workers have room to work, proper sequencing of activities is employed, installed equipment can be placed without interference, etc. Photographs are taken at each step. These photographs are then incorporated into a booklet which is used for indoctrination of field construction personnel at time of actual construction. The crews are able to see, in model form, the step-by-step sequence for accomplishing the task.

Converting nonproductive time to direct work time is only part of the solution to higher productivity. The fact that a worker is in motion is one thing, the speed and efficiency of this motion is another. To improve these also requires management action. Work organization, training incentives, favorable working environment, and similar factors apply here. In addition, the procedures for accomplishing a given task are important. For example, the forms for a wall could be fabricated completely at the place of use, they could be prefabricated adjacent to the place of use and then erected, or commercially available reusable forms could be used. For each choice, the productivity can be expected to be different in terms of square feet of forms placed per hour.

While we should all acknowledge that worker productivity can probably be improved, the estimator for a new construction project would be foolhardy to assume that productivity improvements are forthcoming. The estimator can only assume that the past experience of the firm is representative of what will happen in the future. This is the conservative approach. If management indeed takes action to improve productivity above that assumed by the estimator, the firm will be rewarded by greater profits on that project but the new experience data would alter the base used on future projects being estimated.

Each construction firm's handling of its construction operations is different and for the reasons mentioned above, as well as those yet to be discussed, will result in different general productivity rates. Thus, it is vital that each estimator become completely familiar with the firm he or she works for by observing actual methods employed, by thoroughly analyzing and cataloging field reports, and by working closely with those field personnel who will be responsible for actual field work. From all of this the estimator will develop a data bank, computerized or otherwise, on productivity for various construction operations, and will also develop a feel for the firm which will permit sound judgmental decisions in those areas for which data is not specifically available.

So, let us assume that all modern construction companies are serious about the need to accumulate experience data on productivity. We must now insure that these data are in a form usable for estimating purposes. If the firm conducts all of its construction in a single locality, all data accumulated are more directly usable than is the case if it is a mixture of data from scattered locations around the United States or the world. As will be emphasized in subsequent paragraphs, data must be related to specific areas and specific job conditions. To handle this with least difficulty, the recommended procedure is to maintain base productivity information on jobs located in the vicinity of the home office and then to adjust this information to other areas and abnormal job conditions. Discussion will now proceed to a procedure for doing this. The procedure is that of the author. It is not an exact method, nor could there ever be an exact method to predict productivity. In presenting it, the objective is to outline a reasonable approach to evaluating the effect of the many factors which influence labor productivity. As will be seen, considerable judgment is still required in using the method.

THE AREA PRODUCTIVITY INDEX

Character of the Work Force

It is common knowledge in the construction industry that a work force from New York City will have a different production rate than a similarly sized and experienced work force from Denver, Houston, or Minneapolis. Also, those contractors engaged in international work have learned that work forces recruited from South Korea, Indonesia, Italy, Pakistan, and a host of other nations show significant differences in productivity on the same work. The point is that a blend of many factors gives a given population group a character and quality all of its own. Figure 8-3 is a graphical representation of this. This graph shows variation for three areas in addition to the base area. Referring to the base area first, you will note that productivity varies over a considerable

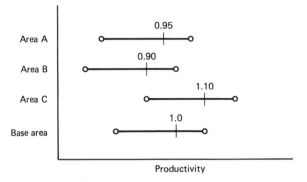

Fig. 8-3   Labor productivity variation.

range as represented by the length of the line between low and high nodes. This range results from the fact that similar work is never accomplished under exactly the same set of site conditions. Weather, congestion, and many other factors yet to be discussed cause work to progress at different rates. On the bar for the base area is a point designated as "1.0"—this point is the normal productivity point for the base area. It is an index for the productivity which has the greatest frequency of occurrence. This is graphically explained in Fig. 8-4. Returning again to Fig. 8-3 and the bars for the other three areas, you note that these bars are offset from that of the base area. These offsets reflect population group productivity differences. Here again each bar has a point with a numerical identification. This identification is an index number that has been derived relative to the base area productivity point with greatest frequency of occurrence. Note that these values can be less than, equal to, or greater than that of the base area as appropriate to indicate population group productivity characteristics. Table 8-2 shows indices that a Houston-based contractor might use.

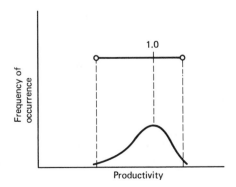

Fig. 8-4   Distribution of productivity.

TABLE 8-2. Productivity Indices for a Houston Contractor

Houston (base)	1.00
Baton Rouge, LA	0.85
Beaumont, TX	0.85
Chicago	0.80
Corpus Christi, TX	1.10
Denver	0.95
Kansas City	0.80
Los Angeles	0.90
New Jersey (north)	0.75
New Jersey (south)	0.80
Ohio Valley	0.90
San Francisco	0.90

Evaluation of these differences becomes a combination of measurement and judgment. If a firm has experience in areas other than its base area, it can possibly compare report data to determine the ratio between productivity in other areas to that of the base area. Some evaluation of productivity might be obtained from potential subcontractors in the area of the work or from other major contractors with experience in the area. Various cost guides available commercially also may prove helpful since they often provide indices of labor costs per unit of work for a number of cities. However, these indices may reflect both crew cost and productivity differences and must be evaluated accordingly.

Actually, there can be two Area Productivity Indices for each location. One of them is for work done under union contract (union shop); the other is for open shop.

Open Shop Versus Union Shop

The differences between open shop and union shop were discussed in Chap. 6. This is a subject area that cannot be avoided in any realistic discussion of productivity since significant differences in productivity can exist between the two options.

A contractor may or may not have a choice between union or open shop operations. In some areas of the country, virtually all qualified craftsmen are union members and essentially all major construction contract work in that area will be union. Contractors electing the open shop approach in such an area face two potential problems. First, they may be able to attract only poorly qualified workers and all aspects of the project will suffer accordingly. Second, they may be subjected to various forms of union reprisals which can affect their progress and costs. In those areas where both open-shop and union-shop operations are legal and feasible, contractors must carefully survey the labor availability, union and nonunion, to determine which method will provide them with the required level of qualified personnel. It is very possible, for example, that other large open-shop contracts in the area will have absorbed the bulk of the qualified nonunion personnel. Assuming that there are equally qualified union and nonunion craftsmen in an area, contractors choosing the open-shop approach should realize greater overall productivity per paid employee-hour than is possible with the union-shop method. The work rules and other conditions peculiar to union projects have the effect of adding nonproductive, but paid, employee-hours to the job so overall average productivity per worker becomes lower. The procedure used to reflect the difference between open and union shop is to use two Area Productivity Indices; one for open shop and one for union shop. The ratio between these two is approximately

$$\frac{\text{Area productivity index for open shop}}{\text{Area productivity index for union shop}} = 1.10$$

The actual ratio is very area sensitive since the level of union activity and its effect on jobs varies tremendously. In some areas there may be no significant difference.

Cataloging the Area Productivity Indices

The ultimate objective of the evaluation discussed above is to develop a catalog of area productivity indices which can be used by the estimating group. Such a catalog will consider both the open- and union-shop conditions. Remember that these indices reflect relative capabilities under the same job conditions. As illustrated in Fig. 8-4, productivity in a given area has a frequency distribution of some form. The reasons for this distribution are the next subjects we must consider.

JOB PRODUCTIVITY ADJUSTMENT FACTORS

In this section a number of conditions will be discussed which affect labor productivity. Each discussion also presents a procedure for determining an adjustment factor for that condition. Ultimately, these adjustment factors will be tabulated and totaled for each project being analyzed and then used in the following equation:

$$\text{Productivity multiplier} = \frac{\text{Area productivity index}}{1 + \sum \text{Adjustment factors}}$$

The adjustment factors suggested in these paragraphs cannot be considered as rules. They are "ball park" factors only and must be adjusted by each contractor as he or she gains experience. Additional conditions not discussed in this chapter may be foreseen which could affect productivity. It is the estimator's job to identify these conditions and assign an adjustment using his or her best judgment.

Adjustment Factors Relating to Project Type

Project Complexity. It is relatively easy to draw a straight line or a perfect circle as compared to a complex curve. Similarly, with construction work it is much easier to handle the commonplace construction tasks as compared to those associated with exotic construction. Estimators normally have considerable data on the ability of their company's crews to form a standard straight foundation wall. But, if this wall is of unusual configuration, that productivity data is completely misleading since much more time must be spent per square foot in building and alignment of unique forms. There also may be features of the project which involve construction operations never performed by the estimator's company. Recall in the construction of the Montreal Olympic complex that contractors had never before encountered epoxy gluing of precast concrete members, no two of which were exactly alike. Such new or one-of-a-kind operations on a construction project will be done with a lower average productivity rate than if done for the second, fifth, or fiftieth time. It is all a matter of the *learning curve* which is based upon the well-known fact that a person's ability to do a given job improves with each repetition and that the time required for each repetition decreases with the number of repetitions.

Much of the information in the estimating group's data base reflects repeated experience. In the course of a past project, there may have been thousands of similar welds or 50 similar concrete pours. The time spent on the first effort in each case is certainly more than that for the last and the overall average productivity would exceed that of the first effort. On the other hand, some of the information in the data base could be that for unique operations, those which were executed once and not repeated again. It should not be too difficult for an estimating group with experience in a firm to know which applies.

In considering the productivity of a crew which will repeat an operation many times in a project, no productivity adjustment would be applied if the data base reflects experienced crews. In other words, for a standard job for the construction firm, there is no adjustment. However, if a new project is extremely complex in comparison to the firm's past experience and/or contains largely one-of-a-kind work packages, an adjustment factor should be applied. The amount of adjustment should

TABLE 8-3.   Design Complexity Adjustment Factors

Project Type	Adjustment Factor
Standard type project for firm.	0.00
Project design complex. Unique work packages.	0.10–0.50

be weighted to reflect the percentage of the project that contains the unusual work packages. The range is given in Table 8-3.

If a type of structure is repeated one or more times on a project and the structures are built in succession rather than concurrently, a contractor can expect greater overall productivity on the second and later structures. As a rough rule-of-thumb, the effort on the first structure, in terms of employee-hours per unit, will be about 10% higher than on following structures.

Quality Assurance/Quality Control.   If the project is subject to quality assurance control (as with nuclear-safety-related construction) all affected work is subject to more frequent and stringent inspection and testing. This causes the workers to be more deliberate and/or frustrated in their work so their productivity is lowered. In addition, the time lost for quality control actions and preparation of QA/QC documentation contributes to more nonproductive time for the workers. An adjustment factor of from 0.10 to 0.20 is appropriate if QA/QC control applies.

Adjustment Factors Relating to Project Location

Weather Factors.   The effect of extreme weather conditions on productivity is easy to visualize. On a hot, humid summer day a worker will be taking frequent breaks to wipe his or her brow or get a cool drink of water and sweat-drenched clothing will cling annoyingly to the body.   Under extreme cold conditions, the worker must have heavy gloves and more layers of clothing, all of which will tend to impede efficiency and he or she frequently will be seeking the shelter of a warming house or bonfire. Actually, a human being can work without degradation of efficiency over a rather broad range of temperature and humidity conditions.   However, once extremes are encountered, rapid degradation can be expected. Figure 8-5 shows a series of curves which represent adjustment factors to be applied to compensate for efficiency degradation due to temperature and humidity. Note that one enters the chart with the temperature (or chill factor in case of colder temperatures) and the relative humidity and then, by interpolation, reads the appropriate adjustment factor.

The factors shown in this figure are applicable to workers performing manual labor without particular exertion (such as electrical installa-

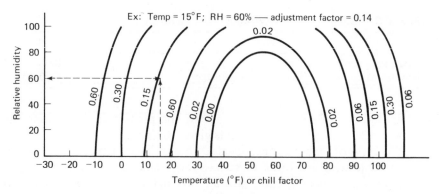

Fig. 8-5   Temperature/humidity adjustment factors.

tion). The factors should be increased by up to 30% if great exertion (lifting heavy materials) or strong mental concentration (arc welding) is required.

Since most projects of any size will involve a number of months or even years to complete, it is important that an estimator apply an adjustment factor for temperature and humidity using a weighted average.

---

example:

A project will be completed during the months of January, February, and March. Adjustment factors and numbers of workers are given in the chart below.

Month	Adj. Factor	Number of Workers	Product
(1)	(2)	(3)	(2) × (3)
Jan.	0.15	10	1.5
Feb.	0.10	16	1.6
Mar.	0.06	25	1.5
		51	4.6

$$\text{Weighted factor} = \frac{(\text{Product})}{(\text{Total worker-months})} = \frac{(4.6)}{(51)} = 0.09$$

If conditions outside the normal (0.0) range are seldom encountered, it would be appropriate to ignore any adjustment. Another cautionary note is in order. The conditions considered are those which will be encountered in the area where work is actually performed. Thus, if a large building under construction is enclosed during the winter in plastic film and space heaters are employed to protect workers inside, there would be no degradation of performance even though outside conditions are extremely cold.

Labor Availability.   If labor is plentiful in an area a contractor will have a better chance of hiring the more skilled individuals in each craft since competition for work will be high.   Also, the workers will tend to be more productive on the job since they know they can be replaced.   The overall productivity will tend to decrease as the size of the work force enlarges on a job, especially if the project begins to absorb all available labor.   These conditions particularly apply to the large contractors with work forces that exceed several hundred.   The small contractor would probably experience less productivity fluctuation during the life of his or her projects but the overall productivity of the small contractor's crews can suffer if he or she is competing for labor with large contractors on nearby projects.   The large contractor often will offer a higher wage, some extra benefit or job continuity that the small contractor cannot provide.   Construction labor loyalty is transitory so such attractions may cause a movement of labor to the larger contractor.   Thus, the small contractor may be left with lower quality workers and frequent worker movement, both conditions lowering overall crew productivity.   To develop an adjustment factor for labor availability, a ratio is first established between size of qualified labor force available and labor force required.   Once this ratio is calculated, an adjustment factor can be obtained from the curve in Fig. 8-6.

Job Site Congestion.   It is no surprise that it is easier to build a building in an open field on the edge of town than it is to sandwich it

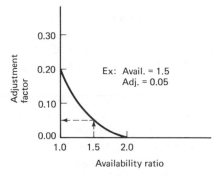

Example: Normal productivity of a crew under excellent conditions of labor availability is 100 units/day. What productivity should be expected if the availability ratio is 1.5?

Fig. 8-6   Labor availability factor.   *Ans:* $\dfrac{1}{1+0.05}$ X (100 units/day) = 95 units/day

TABLE 8-4. Work Space Adjustment Factors

Condition	Adjustment Factor
Adequate Crew Work Space	0.00
Crowded. Approximately one-half space desired.	0.15
Very congested. Approximately one-third space desired.	0.30

between two existing buildings on Main Street. In other words, job site congestion has a considerable effect on productivity since people, machines, and material are competing for use of the same space. This problem can apply to total jobs or to parts of a job. In the case of an industrial plant being built in a rural area, adequate perimeter space is generally available for all administrative and construction activities. But, some structures of the project are complex networks of piping and electrical lines that stagger the imagination. An estimator cannot assume that a welder, pipefitter, or electrician can perform a given task in a confined space at the same rate it could be done in an open space. The adjustment factors for various degrees of congestion are given in Table 8-4.

Adjustments Relating to Project Organization

Overtime. An employee who knows he or she is going to work overtime will reduce his productivity throughout both the normal period of work and the overtime period. This is a psychological effect but nonetheless very real in its effect. A normal work day is 8 hours and and a normal work week is 5 days. Thus, work beyond these normal periods will cause an overall reduction of productivity. The adjustment factor to account for this loss of productivity is found by reference to Fig. 8-7.

An estimator should adjust the average productivity figures for overtime only if that overtime is used heavily. On some projects, the nature of the work being performed calls for considerable scheduled overtime. This is particularly true on projects that are weather sensitive (such as earth moving), on projects in areas of the world with short construction seasons (such as Arctic regions), or on any project with no-later-than completion dates. There are also those projects on which the overtime is only casual (or spot) as, for example, that which would be used to insure that a freshly placed concrete slab is finished before it sets. When overtime is casual, the productivity is not adjusted but an adjustment will be made to the hourly cost of the employee or a lump

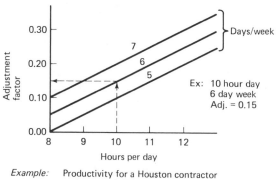

Example: Productivity for a Houston contractor under normal 40 hour workweek conditions is 20 units per day. What will be the productivity in Chicago when working a 6 day week, 10 hours/day?

Ans: From Table 8-2 Chicago index is 0.80
From above, overtime adjustment = 0.15

Productivity multiplier = $\dfrac{0.80}{1 + 0.15}$ = 0.70

Units/day = $\dfrac{20 \text{ units}}{8 \text{ hours}}$ (10 hours)(0.70) = 17.5/day

Fig. 8-7 Overtime adjustment factors.

sum will be added to the estimate to reflect the anticipated added cost of casual overtime (see Chap. 7). Estimators should not necessarily make productivity adjustments in proportion to the anticipated amount of premium pay for the project since, on some projects, premium pay is based upon the time of day or week and not upon whether such time represents working time in excess of 8 hours per day or 40 hours per week. In other words, a worker on a late shift or whose 5-day work week includes Saturday may receive premium pay but the worker is not actually working overtime and the productivity should be considered normal.

Project/Construction Management Organization. In earlier discussion in this chapter, the subject of management actions to improve productivity were discussed. As noted then, the ability of management to efficiently control the project can vary, even within the same firm. Another factor which can affect productivity is the organizational structure employed by the client to administer the overall project. If the various management responsibilities are split, overlapping, and poorly defined among the client, the design element, and the construction element, there are bound to be conflicts, omissions, and delays in construction. These are much less likely with an organizational structure in which responsibility and authority are clearly allocated.

Evaluation of a potential project/construction management structure is difficult. Those construction firms that have been in business a number of years will have served a number of clients and will have built from designs provided by different architect/engineering firms. They will have found that each client and each A/E firm has its own personality, the character of which can affect job progress. Contractors may have had such unfavorable experiences with some clients and A/E firms that their work will no longer be considered. But, overall, this past experience will give the contractor a feel for the general effect of management form on project productivity so that unfavorable situations can be flagged in advance and appropriate adjustments made.

Following are some specific management organization features which should be closely examined and evaluated for their effect on productivity.

1. *Accessibility of Designers.* It must be assumed that there will be change orders during the course of a project due to errors, deletions, or additions. If the responsible design agency maintains a staff at the project site or very nearby with change authority, responsiveness will be greater than would be the case if all authority is located in a distant city. Any lack of responsiveness on the part of the design group can cause project delays with attendant impact on productivity.

2. *First Time Experience for Project/Construction Management Group on This Type Project.* As discussed under Project Complexity there is a learning curve that applies to all repeat operations. The learning curve also applies to management. If this is a first time experience on this type of project for a management group they will do a lot of expensive learning. The added expense will come about through the lowered productivity of a confused and frustrated work force that is subjected to poor decisions or indecision. An experienced group will have fully developed procedures that have been refined through experience. In fact, the larger firms will have all such procedures in writing.

Table 8-5 gives suggested adjustment factors to be used based upon the project/construction management situation.

TABLE 8-5.   Management Experience Adjustment Factors

Situation	Adjustment Factor
Well-coordinated, experienced management team including client, A/E, and contractor.	0.00
Possibility of difficulty among client, A/E, and contractor	0.05–0.20
First time experience for client or A/E project management elements or construction management team.	0.20–0.50

Control of Project Schedule. It is assumed that any contractor, large or small, will use planning and scheduling procedures available and appropriate to the level of operation to assure the most efficient integration of labor, equipment, and materials. Such planning is definitely possible if the project has been fully defined, adequate labor and equipment are available, and material and installed equipment deliveries can be assured when needed. For many projects, one or more of these conditions is not assured at the start of construction. On turnkey projects engineering may find it hard to keep up with construction. This causes designs to become rushed with attendant errors, change orders, and rework. Rework is extremely demoralizing to crews and crew productivity suffers accordingly. Procurement difficulties are another problem. Very often, key items of construction or installed equipment or construction materials will not have off-the-shelf availability or firm delivery dates. Delays of any of these items can delay an entire work force. The problem becomes compounded if a contractor is working against an established completion date because he or she can be forced into overtime conditions and/or the addition of crews to the job with attendant congestion and subsequent loss of productivity.

A number of scheduling problems can be expected in any project. In fact the base data used by the estimating group is for average performance during normal conditions and average performance implies that some problems were encountered since better performances have been registered. Thus, a productivity adjustment factor to reflect lack of project scheduling control should be introduced only when unusual circumstances are expected which can throw the scheduling off more than that of the average project. The adjustment factor selected is even more judgmental than some of the previous factors. Ranges are given in Table 8-6.

TABLE 8-6.   Scheduling Uncertainty Adjustment Factors

Scheduling Conditions	Adjustment Factor
Average project, average problems, scheduling constraints reasonably predictable.	0.00
Some possibility of problems due to poorly defined work, materials shortages, labor strikes, deadlines, or procurement lead times.	0.01–0.50

Other Adjustments

The conditions and suggested adjustment factors described thus far are among the more common possibilities. Additional conditions

TABLE 8-7. Adjustment Factor Rules-of-Thumb

Condition	Adjustment Factor
The condition, by itself, would tend to cut productivity in half	1.0
The condition, by itself, would tend to cut productivity 25%	0.3
The condition, by itself, would tend to cut productivity by 10%	0.1

affecting labor productivity may certainly exist. Determination of an appropriate adjustment factor can be done using the guidelines of Table 8-7.

DEVELOPING THE PRODUCTIVITY MULTIPLIER

In the introductory section of this chapter some statistics were given on productivity experiences at nuclear power plants on similar work. To illustrate the use of the indices and factors discussed in this chapter and to show how productivity rates as different as those documented in Table 8-1 can be possible, the following example is given:

example:

Assume Area A is the base area. It is located in an area where open shop operations are commonplace. A project will be built in Area A. All conditions concerning this contract are favorable.

Assume a similar project is to be built in Area B. It will be a union job, labor availability is marginal (estimated at a ratio of 1 to 1.1), union contracts are subject to renegotiation every 2 years, and strikes at contract time have been the rule. The clients have decided to handle the project management function themselves even though they have not had previous experience in this role. The design is being done by an A/E who will also handle construction management. All work will be union subcontract. Weather will be a factor about 25% of the time when average temperature will be 25 degrees with a 50% relative humidity. The Area Productivity Index for Area B is 0.85 for union contract. (See Table 8-8 for adjustments.)

$$\text{Productivity multiplier, Area B} = \frac{\text{Area productivity index}}{1 + \text{Adjustment factors}}$$

$$= \frac{0.85}{1 + 0.575} = 0.54$$

TABLE 8-8. Summary of Adjustments

	Area A	Area B
Area Productivity Index	1.00	0.85
Adjustments:		
Project complexity	NA	NA
Quality assurance	NA	NA
Weather	NA	0.015
Labor availability	NA	0.16
Project congestion	NA	NA
Overtime	NA	NA
Management organization	NA	0.20
Control of schedule	NA	0.20
Other	NA	NA
Total Adjustments	None	0.575

In this example the ratio of productivity in Area A to that of Area B is almost 2 to 1; a situation that can be expected under conditions outlined. The estimating group would adjust their base productivity data by applying a multiplier of 0.54 in their calculations.

You may have noted that the formula for determining the productivity multiplier is of a form which reduces the effect of individual adjustment factors in the presence of other adjustment factors. For example, assume only a single adjustment factor of 0.1 is applicable against an area productivity index of 1.0. The productivity multiplier would be

$$\text{Productivity multiplier} = \frac{1.0}{1 + 0.1} = 0.91$$

This is a degradation of 0.09. Now, consider the effect of an additional adjustment factor of 0.1.

$$\text{Productivity multiplier} = \frac{1.0}{1 + 0.1 + 0.1} = 0.83$$

The effect of the second adjustment factor is only 0.08 more, not 0.09. This condition is realistic in that factors affecting productivity become interrelated so their effect is not purely additive.

CONSIDERING PRODUCTIVITY VARIATION IN A MAJOR ESTIMATE

It was stated earlier that productivity of labor is the major variable in the estimating equation; it is not impossible to experience variations

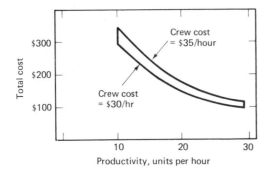

Fig. 8-8   Range of costs due to variables.

from low to high of 1 to 3. Labor cost per hour can also vary, but this tends to be more predictable since wage scales are known at time of bidding. An element of uncertainty does exist on long projects because the number and extent of wage increases can only be estimated. Figure 8-8 illustrates the challenge these variations force on the estimator. In this figure it is assumed that a work item consists of 100 units, productivity can vary from 10 to 30 units per hour for our crew, and the crew cost can vary from $30 to $35 per hour. These variations create a zone of possible cost of the work that goes from a low of $100 to a high of $350. Within this range the estimator must select a figure which covers the cost of doing the work without being so high as to make the bid noncompetitive.

Chapter 16 will present the subject of statistical bidding as part of the overall procedure for developing an estimate for a major project. In this process the estimator works within a range of productivity, the lower value in this range representing the lowest possible productivity he or she can envision at the project site if almost everything goes wrong. On the other hand, the high value will be the value of productivity if everything goes right. Somewhere in between the estimator will pick a "target" value which represents his or her best estimate of actual productivity. Then by means of a computer simulation involving all variables in the project, including productivity of each craft, a statistical distribution of probable costs is determined which enables the estimators and managers to determine bid prices and their associated probability of being overrun or underrun. In this chapter you have been given productivity guidelines which will better enable you to select these values.

SUMMARY

Chapter 1, in its discussion of the construction of the Olympics Complex in Montreal, noted that the cost per seat in the main stadium reached a level of about $13,000 while that of the New Orleans Super-dome was less than $2500 per seat. Certainly, productivity problems of the type discussed in this chapter were heavy contributors to this escalation.

The construction cost estimator must realize that productivity is the real variable in the labor cost equation. Estimating productivity calls upon an estimator's best talents since he or she must simulta-neously evaluate the effect of many interrelated factors. Inherent in this evaluation is considerable judgment which can be gained only through experience and research.

REVIEW

Exercises

1. Assume that your company is based in Houston, Texas, and that the Area Pro-ductivity Indices given in Table 8-2 are used by your firm. Develop the pro-ductivity multiplier you will recommend under each of the following sets of conditions:
   a. The project being bid is in Denver, Colorado. Labor availability ratio for carpenters is expected to be no better than 1.25. The project is located in a rather tight area so that your carpenters will have only about half the space they normally like to have.
   b. The project is located in Corpus Christi, Texas. The only problem you forsee involves the client and his engineer. The client is the "nervous" type and the engineering firm has had almost no prior experience on this type of design.

2. On a building project you plan to utilize a crew of carpenters to place 5000 SF of forms for one of the pours. The crew cost per hour is $35. The crew pro-ductivity rate is 335 square feet per day on this type of work under normal conditions.
   a. What is the labor component of cost for installing these forms under normal conditions?
   b. Unfortunately, heavy rains have delayed the form work. Since slippage of this work will delay the entire project you decide to work the crew 3 hours overtime each weekday and to work them on Saturdays for 11 hours. Pre-mium rate for overtime and weekend work is $1\frac{1}{2}$. Assuming the first day of work is a Monday and that there will be no further weather interruptions, when will the forming be complete?
   c. What will be the cost of the work under the situation described in part b?
   d. Another approach for handling the problem described in part b is to double the crew. However, the regular crew size is the most efficient for the work so doubling the crew has the effect of giving the workers only half the work space they need to operate efficiently. If the choice is made to double the crew, what is the relative efficiency of this option compared to using a single crew?

e. With the situation described in part d, and assuming that the first day of work is a Monday, when will the forming be complete if only a standard 40-hour week is used?

f. What are the costs of the option described in parts d and e?

g. If you decide to use double crews, a 6-day workweek, and an 11-hour day, what is the relative efficiency of this option in comparison to a single crew working regular time?

h. In part g, what will be the total cost to complete the forming?

3. You have been assigned as lead estimator for a university academic building project in Fargo, North Dakota. Your research indicates that the Area Productivity Index for Fargo should be approximately 1.10 in comparison to your base area. You have examined the contract documents and visited the site and have identified the conditions listed below. Determine an appropriate overall productivity multiplier.

a. Project takes advantage of preengineered components and standard modules throughout.

b. Contract operations can begin on June 1. The building must be ready for occupancy within 26 months.

c. Temperatures in Fargo range from an average high of $90°F$ in summer to an average low of $-34°F$ in winter. Winter wind chill factors often exceed $-50°F$. Freezing conditions occur as early as mid-September and as late as the end of May. Crew sizes during the months from October through March of each year will be only about one-third those during the remaining months.

d. Normal scheduling should permit completion of the project in 26 months.

e. The project is located near the center of the university. Your operations cannot disrupt normal academic activity and no streets can be reserved for construction access. You must erect a safety barrier around the project that is no farther than 30 feet from the proposed building perimeter and all operations must be contained within this barrier. A materials storage area is available approximately 500 yards from the project.

f. This probably will be the largest construction project underway in the Fargo area during the period based upon known requests for construction permits. Contact with state and local labor offices indicates that adequate labor will be available but not plentiful. Your estimate is that the overall availability ratio is about 1.2.

g. Your firm has had no prior experience with the A/E firm that designed the structure. However, you have had more than usual difficulty in reading and interpreting the contract documents. Inquiries to contractors who have experience with this A/E state that he normally assigns a full-time inspector to a job of this size and will "hold your feet to the fire" in his role as client representative.

Questions in Review

1. The average American worker spends about _____ of his or her time in direct work.

2. An approximate upper limit for direct work time as compared to total work time is _____ .

3. Why does union labor productivity tend to be lower than that of nonunion labor on an employee-hours per unit of work basis?

4. What does the base data on productivity represent?

5. In what way does overtime affect productivity? Cost?

6. To what extent can productivity vary on the same type of work?

Discussion Questions

1. As a potential construction manager, what might you do to ensure maximum productivity within your work force?

2. How would you determine the efficiency of one of your work crews?

3. How are job safety programs and productivity related?

4. What motivates a worker (or you) to be more productive?

5. Do you think a construction firm should maintain productivity data on supervisors? How would you use it?

6. Productivity of the American worker is said to be declining. Why is this true? What can be done about it?

7. A contractor has bid on a project assuming a certain major operation could be completed during the months of April and May when weather conditions were favorable as to temperature and humidity. However, the client causes a delay so that this operation is held off until July and August when temperature and humidity conditions are very unfavorable and productivity suffers. Should this be cause for a claim against the client? If so, in what way should the contractor establish a basis for the claim?

INTRODUCTION
TO CONSTRUCTION
EQUIPMENT

9

INTRODUCTION

Another resource to be accounted for by the estimator is construction equipment. This is the equipment utilized directly or indirectly in the installation of materials and permanent equipment in the structures to be turned over to the client. At the time an estimate is being prepared it is unlikely that the estimator will have an exact listing of the items of construction equipment that will be used on the job; the estimator must know enough about the work in question to make appropriate assumptions concerning types and amount of use of construction equipment.

The typical project begins with the civil construction work. Site clearing, development of access, shaping of the site, and foundation excavation all involve equipment in the earthmoving category. Civil construction continues with construction of foundations. A foundation may be on pilings, drilled piers, or placed directly on subgrade strata. Whatever its bearing, it will normally contain reinforced concrete. Above the foundation comes the structural frame, walls, and any floors or decks. These later stages of civil construction require hoisting equipment, materials handling equipment, concrete placement equipment, and any number of specialized items such as welding machines, air compressors, generators, and power tools. As civil construction winds down mechanical, electrical, instrumentation, and other specialized work begins as permanent equipment and systems are installed and finish work is accomplished. Again, various types of hoisting and materials handling equipment will be needed in addition to the specialized tools of each craft. In late stages of construction, the site must be cleaned up and trimmed before being turned over to the client. Trucks, loading equipment, and other items will be needed for this final effort. In this chapter the more common types of construction equipment will be described.

EARTHMOVING EQUIPMENT

Crawler Tractors

The basic unit is a track-mounted tractor, usually diesel engine driven, to which may be added one or more attachments. Tractors come in a complete range of sizes permitting selection of power to best fit the needs of a given job. Common configurations, (shown in Fig. 9-1) of the crawler tractor are

Bulldozer.   In this configuration, a blade is mounted on the front of the tractor for purposes of moving soil by pushing straight forward. The blade can be raised or lowered but not rotated about a vertical axis. The bulldozer is suitable for moving soil for short distances only, up to about 150 ft [50 meters].

Angledozer.   The blade mounted on this tractor can be rotated about a vertical axis as well as raised and lowered to permit sidecasting of material.

Treefelling.   The blade on this tractor is a specially designed angle-blade with a serrated cutting edge which permits the cutting of trees while also pushing them over.

Pushdozer.   This dozer looks much like a bulldozer except that the blade is shorter and specially reinforced for pushing scrapers as a

Fig. 9-1 Crawler tractor operations. (Courtesy Caterpillar Tractor Co.)

means of helping them load. It approaches the scraper from behind and engages a push block on the rear of the scraper with the push blade. A bulldozer can be used for pushing scrapers if a pushdozer is not available.

Winch. The winch attachment, normally mounted on the rear of the tractor, is in addition to one of the attachments described above to provide winching power for towing stalled equipment, removing root systems, and other heavy-duty pulling. Normally, only a part of the dozer fleet is equipped with winches.

Ripper. The ripper attachment is a device mounted behind the tractor and contains one or more large teeth which point downward and forward. In use the ripper is like a plow and is used to break up soft rock, asphaltic pavement, thin unreinforced concrete, etc., for movement by other means.

Rubber-Tired Tractors

This piece of equipment is similar to the crawler tractor except it is mounted on four rubber tires. It can be equipped like the crawler tractor or used without attachments for towing purposes. Being wheel-mounted, it is less useful than a crawler tractor on those operations requiring great traction. However, the rubber tires give it greater speed making it the better choice for other operations.

Earth Scrapers

Scrapers consist of a prime-mover and a large towed bowl or hopper. They are used for the loading, transport, and spreading of soil. The scraper approaches the lane to be cut and the bowl is lowered so that its cutting edge digs into the material. Using the power of its prime-mover and a pushdozer, if needed, the scraper moves foward cutting the soil and filling the bowl. When full, the bowl is raised and the scraper travels to the fill area. Here the bowl is lowered slightly, a front gate on the bowl opened, and the soil in the bowl is ejected forward and out of the bowl so that it is spread in a fairly uniform layer as the scraper moves forward. Most modern scrapers are rubber-tired and the engine section and bowl section are integrated into a single unit. Occasionally, one will still find a separate scraper mounted on rubber wheels which can be towed by a crawler tractor. The smaller conventional scrapers today contain a single diesel engine that is mounted on an engine section with either one or two axles. Larger scrapers contain a second power unit mounted on the bowl section which drives the bowl section wheels. When cutting, scrapers are normally assisted by the pushdozers. One pushdozer can service a number of scrapers. Scrapers are shown in Fig. 9-2.

There are designs of scrapers intended to eliminate the need for pushdozers. The most common is the elevating scraper. It looks very similar to a conventional scraper except that a ladder elevator is mounted in front of the bowl to move the material up and back into the bowl as the scraper moves forward. This action reduces traction requirements for loading. Another configuration which eliminates pushdozers is the push-pull arrangement. This requires teams of two scrapers, each having a special ring and hook attachment which enables the scrapers to couple and uncouple from each other. In operation, the scrapers couple up elephant fashion, and the rear scraper pushes the front one while it loads, after which the front scraper tows the rear scraper as it loads. They then uncouple and travel independently to the fill area.

Rubber-tired scrapers are capable of haul speeds up to 35 mph and are excellent for earthmoving up to $1\frac{1}{2}$ miles. They require good haul roads and excellent traction for economical operation. A towed scraper can operate under relatively poor traction conditions but its maximum speed is about 6 mph.

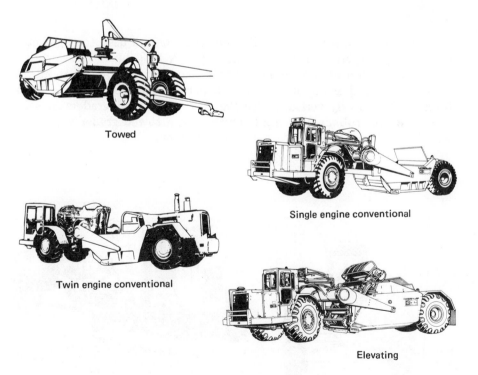

Towed

Single engine conventional

Twin engine conventional

Elevating

Fig. 9-2   Scraper configurations. (Courtesy Caterpillar Tractor Co.)

Highway Trucks

These are the conventional dump trucks in most cases, although there are some towed bottom or end dump trailers that can be included in this category. Highway trucks are used for hauls beyond the economic range of rubber-tired scrapers, or where use of public roads in the hauling dictates use of hauling equipment licensed for road use.

Off-Highway Trucks

These are extremely heavy-duty, high-capacity trucks used for hauling operations where particularly durable equipment is required (as in quarries) or where the haul roads are reserved for construction use and the exterior dimensions (up to 16-ft wide) of the equipment pose no safety problem to other traffic. Standard models are available with capacities as high as 80 tons and speeds in excess of 30 mph and special order trucks of larger capacity have been produced. These trucks are particularly useful in conjunction with high-capacity power shovels for high-volume, high-speed earthmoving operations.

Graders

These items, sometimes called *maintainers*, are used for light cutting, spreading, and finish work on soils. A long blade is suspended about midsection of the frame and this blade can be rotated about all axes and shifted from side to side to permit work on surfaces from horizontal to nearly vertical. See Fig. 9-3. Almost all graders are self-propelled and rubber-tired although there is a towed grader with steel wheels which can be used if traction is a problem.

Fig. 9-3   Grader. (Courtesy Caterpillar Tractor Co.)

Rubber-Tired Loaders

These vehicles are often called *front loaders*. They consist of a specially designed rubber-tired tractor with a bucket mounted at the front. The bucket can be raised to an overhead position or lowered a few inches below the surface on which the loader wheels rest. The bucket also can be rotated on its long axis in order to control the material in the bucket as it is loaded, raised, and dumped. Loaders are used primarily in stockpile operations for the loading of materials into trucks or hoppers but can be used for transport of materials around a work site, light dozing, or for general lifting of materials. They come in many sizes of buckets from a fraction of a cubic yard to several cubic yards. The frame can be either rigid or articulated. Rigid frame models may have either front or rear wheel steering, while articulated models are steered hydraulically by relative turning of the two body sections about the vertical axle joining them. See Fig. 9-4.

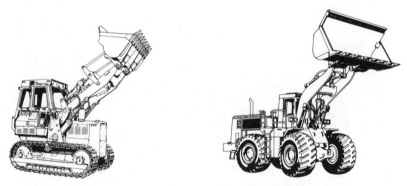

Fig. 9-4   Rubber-tired and crawler loaders.   (Courtesy Caterpillar Tractor Co.)

Crawler Loaders

This type of loader has a rigid frame and is mounted on tracks. Otherwise, it has the same function as the rubber-tired model but is the only type of loader that can be used under conditions of poor traction.

Compactors

Compactors are used to consolidate soils in fills, road bases, etc. Since the method of achieving compaction in a particular soil depends

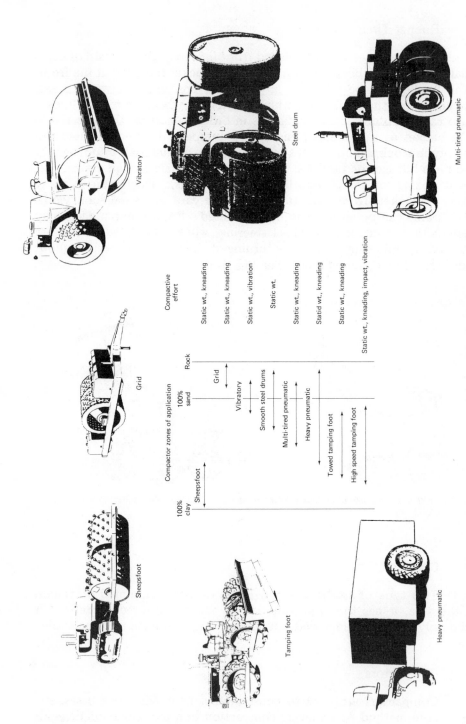

Vibratory

Steel drum

Multi-tired pneumatic

Grid

Sheepsfoot

Tamping foot

Heavy pneumatic

Compactor zones of application

Compactive effort

Rock

100% sand

100% clay

Grid — Static wt., kneading

Vibratory — Static wt., kneading

Smooth steel drums — Static wt., vibration

Multi-tired pneumatic — Static wt.

Heavy pneumatic — Static wt., kneading

Sheepsfoot — Statid wt., kneading

Towed tamping foot — Static wt., kneading

High speed tamping foot — Static wt., kneading, impact, vibration

Fig. 9.5   Compaction equipment and applications.   (Courtesy Caterpilar Tractor Co.)

on that soil's characteristics, many compactor designs are available enabling a contractor to choose that which will best meet the conditions of the contract. Compactors may be either self-propelled or towed, the towing being handled by a crawler or rubber-tired tractor. The most common types of compactor rollers and their applications are described below and illustrated in Fig. 9-5.

Sheepsfoot Roller. This type of roller consists of one or more hollow drums on whose surface are mounted many teeth or feet. Each foot is roughly equivalent to a cylindrical or rectangular rod whose end gives several square inches of contact with the soil. As the roller is pulled, these feet punch into the soil and compact it. Through multiple passes, complete compaction can be obtained. The sheepsfoot roller is suitable for compacting very cohesive soil in lifts of about 6 inches [15 cm] or less. The rollers can be filled with water or sand to get whatever weight is desired.

Tamping Foot Roller. The overall design of this type of roller is similar to that of a sheepsfoot roller except the feet or pads are much larger in surface area. This difference gives them less ability in compacting highly cohesive soils than the sheepsfoot roller but they are still excellent for low to medium cohesive materials.

Grid Roller. With this type of roller, the outer surface of the drum is a heavy duty grid. In passing over the material it combines overall weight with a crushing action to achieve compaction. Ballast boxes permit the addition of weight to the roller. The grid roller is the best compactor for gravels.

Steel Drum Roller. These rollers are self-propelled and contain two or more steel wheels. One form has these wheels in tandem although a three wheel version similar to a farm tractor is used. Steel wheel rollers are used for finish compaction of thin layers of sand, gravel, and sand-gravel-clay combinations used in base courses. It also is used for asphaltic concrete pavement work. Most contain a water tank which adds weight or which permits use of water spray during compaction.

Heavy Pneumatic. These compactors, which are usually towed, are basically huge rectangular steel boxes mounted on four or more large rubber tires. Their tanks can be filled with ballast to give them the total weight needed for the compactive effort desired. These compactors can weigh over 50 tons when ballasted and are used for compaction of moderately cohesive base materials. Their size permits compaction of lifts as much as 12 inches thick.

Vibrating Rollers. Vibration is an excellent means of compacting sands and this type of roller is designed to use this method. It is either a steel wheel, grid, or sheepsfoot roller with an auxiliary device to provide vibration to the roller.

Multitired Pneumatic. This type of roller contains two banks of rubber-tired wheels each about the size of a large automobile wheel but with smooth tread. The wheels are mounted singly or in pairs in a line on independent axles. This arrangement of the wheels imparts a kneading action as well as vertical pressure on the soil being compacted. They are good for sands, gravels, or those materials in combination with some cohesive material. The body of the roller is a box into which ballast can be loaded to provide weight needed.

Crane-Body-Mounted Excavating Equipment

A number of pieces of excavating equipment are mounted on crane bodies. The crane body, in turn, may be mounted either on tracks, wheels, or a special-purpose truck chassis. See Fig. 9-6.

Power Shovel. Power-shovel digging action is accomplished by forcing the cutting edge of a bucket into the material by both lifting and crowding (pushing forward). When the bucket is loaded the shovel rotates to dump the load into a waiting truck or onto a stockpile. This shovel is particularly good for excavation from an embankment face. Bucket sizes range from less than a cubic yard to several cubic yards for ordinary work. Special-purpose machines with capacities of 50 or more cubic yards have been specially fabricated.

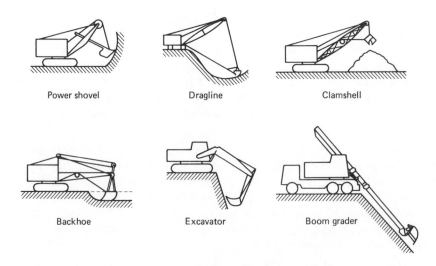

Fig. 9-6   Crane-body-mounted excavating equipment.

Dragline. A long truss-boom is used in this configuration, the boom being controlled with cable. A box-like bucket that is open on top and at the end facing the machine is rigged with another cable line through a sheave at the end of the boom. Cutting teeth are along the bottom edge of the open end. The operator throws the bucket by rotating the crane and releasing the bucket much like sidecasting a fishing lure. A cable mounted on the open end of the bucket is then reeled in by the operator causing the bucket to scrape over the surface and load material. When loaded, the bucket is raised, the crane rotated, and the bucket dumped. The dragline is excellent for digging operations below the level of the ground on which the dragline is located and is especially useful for excavation underwater or in areas with unstable soil conditions.

Clamshell. The clamshell attachment, like the dragline, is rigged on a crane boom. The clamshell bucket itself is like a pair of suspended jaws, hinged at the top, that can be opened or closed. In operation, the clamshell is raised over the material to be loaded or excavated. Then, with the jaws open, it is dropped on the material. By means of a cable, the jaws are then closed and the entire bucket lifted. The crane is then rotated and the load dumped. The clamshell is excellent for loading of loose materials, such as in a stockpile, and can be used for the same work as a dragline except that its reach is much less and it does not have the same digging power.

Backhoe. The backhoe looks like a power shovel with the bucket reversed so that digging is accomplished as the bucket is pulled toward the crane body. The backhoe is used for trenching work.

Excavator. The excavator is a form of backhoe. However, an excavator is permanently configured as such while a backhoe is but one of a number of attachments that can be attached to a basic crane body. The excavator is hydraulically operated and is excellent for hard digging conditions.

Boom Grader. The boom grader can be considered a specialty backhoe. It has a boom which the operator can extend or contract, raise or lower, or rotate. At the end of the boom is a wide bucket with a cutting blade that faces toward the crane body. This bucket can be rotated within limits about all the axes. The boom grader is used for finish cutting on surfaces such as ditch slopes.

Water Distributor

Water distributors are truck-mounted or towed water tanks with a spray bar. The largest versions are wheel mounted on a chassis similar to that of a scraper and are towed by a large rubber-tired tractor. Water distributors are used for sprinkling of water in fill areas being compacted and for dust control on haul roads serving a project.

Ditching Machines

Ditching machines are of two basic types. The *rotary* ditching machine travels on tracks which straddle the line of a ditch to be dug. To the rear of the machine is a large wheel with digging buckets attached to its periphery. In use, the wheel is caused to rotate and is lowered to the desired depth as it cuts. Material is wasted to the side. The *ladder* style of ditcher has digging buckets attached to an endless chain mounted over a guide frame.

CONCRETE PLACEMENT EQUIPMENT

The sequence of operations for the placement of reinforced concrete starts with base preparation and proceeds through forming, reinforcing steel installation, concrete pouring, concrete curing, and concrete finishing. For slab pours, the finishing operation precedes the curing; for walls, finishing follows curing and form removal. Preparation of a base, if applicable, is handled by earthmoving equipment or hand tools. The forming and resteel operations are often totally accomplished with hand tools although some hoisting equipment may be used to move materials. Finishing operations are handled with a variety of small tools, either hand or motor driven. The operation requiring application of major items of equipment is the pouring of the concrete.

Concrete may be supplied from commercial ready-mix sources or produced on site with a batch plant. If commercial ready-mix is used the cost of the concrete includes all costs of mixing and delivery; if an on-site batch plant is used, the estimator must account for the costs of plant erection, use, and demobilization, as well as remembering to include required batch trucks in the equipment fleet.

The easiest method of placing concrete is to chute it directly into the forms from the delivery truck. This is possible up to distances of about 20 ft. For greater reaches, several options are possible. A very common approach is to use a crane with a concrete bucket attachment. The bucket is loaded from a delivery truck and the bucket moved to point of placement by the crane. Another option is to use concrete buggies, either hand-powered or motorized. The buggies are filled by the delivery truck and the buggies then moved over ramps to point of placement. For very high pours, as in buildings, concrete elevators are available to move the concrete vertically.

Pumping of concrete is an excellent option for large pours at virtually any location. Pumps are available which are skid, trailer, or truck-mounted. These pumps connect with piping sections which are assembled in the field. Specially designed booms organic to the pump or pedestal-mounted for remote locations can be used to place concrete over a wide area from a single delivery point. See Fig. 9-7.

Conveyors are another means of delivering concrete from a central receiving point to the point of placement. Conveyors can be angled as much as 30 degrees vertically and any number can be arranged in series. They do not provide as much flexibility as is possible with pumping but are a better choice if concrete deliveries are intermittent or if difficulties may be encountered in obtaining a proper mix for pumping. Conveyors come mounted on dollies, on rubber-tired crane bodies (creter-cranes), or on pedestals. Pumping and conveying of concrete has largely replaced the use of cranes, buckets, and elevators on major pours.

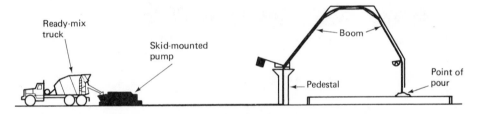

Fig. 9-7   Concrete pumping using a pedestal-mounted boom.

HOISTING EQUIPMENT

Many tons of materials and permanent equipment must be handled during the course of a construction project—it must be moved both horizontally and vertically. The equipment market provides many styles of equipment to handle these items.

Cranes

Some type of crane is found on almost every project. Some are cable controlled, others hydraulic. They are mounted on tracks, wheels, truck chassis, pedestals, and ringers. Through use of boom extensions, jib booms, and derrick booms, a great variety of boom configurations are possible to meet the needs of a job. Some of these are illustrated in Fig. 9-8. Since cranes are very expensive items of equipment it is essential that the job be carefully planned so that needed cranes are there and efficiently utilized.

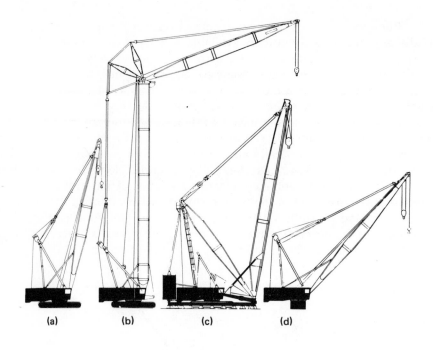

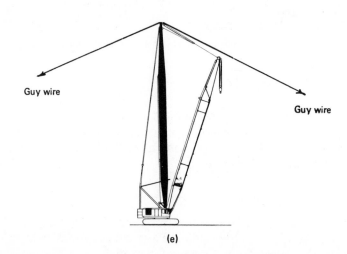

Guy wire

Guy wire

(e)

Fig. 9-8 Crane configurations: (a) conventional lift crane; (b) tower crane; (c) ringer crane; (d) pedestal crane; (e) crawler crane converted to fixed leg derrick; (f) crawler crane with dolly-mounted counterweight; (g) hydraulic rubber-tired crane; and (h) climbing tower crane. (Parts a, b, c, and d courtesy of Manitowoc Engineering Co.; parts e and f courtesy of Amhoist, American Hoist & Derrick Co.; part g courtesy of Galion Manufacturing Co.)

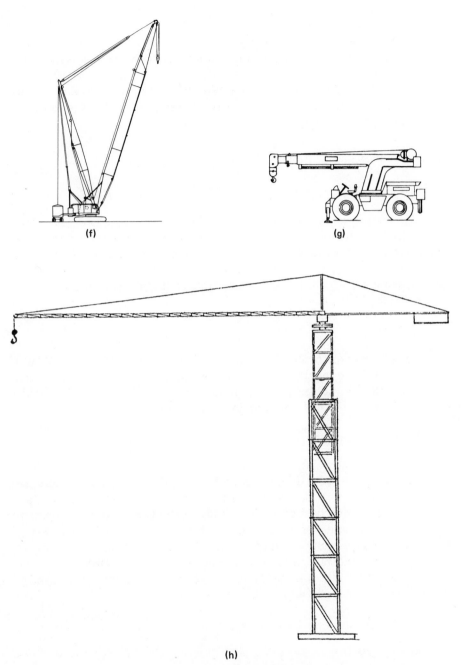

(f)

(g)

(h)

Fig. 9-8 (*Continued*)

Forklifts

Forklift designs are available for use inside warehouses, on hard surface roads, or on rough terrain. Those designed for inside work are normally battery powered; for outside work a combustion engine is used. Forklifts are intended to handle pallets or bundles of materials and can move them either horizontally or vertically within the limits of the machine.

SPECIAL PURPOSE EQUIPMENT

Air Compressors

Air compressors are used to provide air for paving breakers, concrete vibrators, hand compactors, and other pneumatic tools. Most are trailer-mounted although truck- and skid-mounted versions exist. Air compressors are rated in cubic feet of air per minute. Both electric and diesel engine driven models are manufactured and capacities range from about 80 cfm to over 600 cfm.

Generators

Electric generators may be truck-, trailer-, or skid-mounted and come in sizes from less than 5 kW to more than 250 kW. They are used to power electric tools, provide illumination, and otherwise provide electric power if it is not commercially available.

Welding Equipment

Construction welding is normally accomplished by arc welding. Processes used include shielded metal arc (SMAW), tungsten-inert gas (TIG), or metal-inert gas (MIG). All three processes utilize a welding machine that converts source alternating current into ac or dc current of various amperage and voltage combinations for welding. Many welding machines include a motor-generator unit to produce the source current. MIG and TIG processes require an inert gas source to provide required weld puddle shielding gas. Oxyacetylene equipment may also be on site for metal cutting or light welding tasks.

Piledrivers and Pier Drilling Equipment

Many structures have piles or drilled piers for foundations. Piles are driven with piledrivers which are specially configured crane bodies. Piledriving "leads" attached to the front of a crane body position the pile for driving and act as guides for the driving hammer. Piles installed directly include wood, prestressed concrete, or various rolled steel sections. A cast-in-place pile is made by driving a thin corrugated shell into the ground with a mandrel insert, removing the mandrel, and then filling the shell with concrete.

A pier drilling auger can also be mounted on the front of a crane body. This auger is capable of drilling a cylindrical hole in highly cohesive soils and reaming out a bell shaped footing at the bottom of the hole. Reinforcing steel is then positioned in the hole and the hole filled with concrete to create a cast-in-place reinforced concrete pier.

Rock Production Equipment

Crushed rock may originate in a quarry or a river-run source. If from a quarry, the rock must be initially broken away by blasting. To do this, small diameter shafts are drilled into the rock using wagon drills which are powered by air compressors. These shafts, which are drilled in a definite pattern to control the rock break, are then filled with explosives for blasting. The blasted rock is then moved to a primary crushing unit which breaks the large pieces into pieces 5 in. or less in size. These are then passed through successive crushing and screening stages to produce aggregates of the desired size distribution. Washing equipment may be part of the setup to remove foreign materials or fines.

A river-run source contains naturally produced rock and boulders and is usually found along the bottom and banks of a fast-moving stream near the base of a mountain. Similar deposits are found in areas near the boundaries of past glaciers. This rock is removed directly by excavating equipment and fed into crushers. Washing equipment is normally required since the rock source is usually dirty.

Paving Equipment

A wide variety of equipment is available for asphaltic concrete and regular concrete pavement work. This equipment is decribed in the Base Courses and Paving section of Chap. 13.

ADMINISTRATIVE VEHICLES

Pickup Trucks/Carryalls

These vehicles, generally assigned to supervisory personnel, handle both personnel transport and light hauling requirements associated with a project.

Vans

These vehicles come in personnel, cargo, or special-purpose configurations. On some projects, vans are used for the transport of the workforce from a central gathering point to the project. Cargo configurations may be used for general cargo transport or as portable tool and supply rooms. A typical special-purpose configuration is an ambulance.

Lubricators

These items, either truck or trailer mounted, are used to provide lubrication service to construction equipment at locations other than the central motor pool. Custom designed lubricators are available or their equivalent can be assembled by the contractor.

Fuel Trucks

On large jobs a central refueling facility will be maintained but there will be items of equipment, such as crawler cranes, which require servicing and fuel where they operate. This requires the use of truck- or trailer-mounted fuel tanks and pumps.

Contact Maintenance Trucks

These trucks are maintenance shops on wheels and provide transport for mechanics, tools, and parts to locations where maintenance service is required.

Scooters

On very large projects, small three-wheel scooters may be used for messenger service around the project site. Being small, manuverable, and economical, they are ideal for this service.

Tractor-Trailers

This combination is composed of a truck prime-mover (tractor) and a towed trailer. The trailer may be a *lowboy* which has a low flat bed with supporting axles and wheels to its rear. Loading is accomplished over the end of the trailer by driving up steel ramps. Another trailer version is the *tilt-bed*. On this, the trailer bed is also flat but the supporting wheels and axles are further forward. The bed is hinged to the trailer frame aft of its midpoint so that a piece of equipment, in unloading, will cause the bed to tip downward and become its own ramp. Other specially designed trailers are available for transport of beams, pipes, and other items.

Stake-Bed Trucks

These are general cargo vehicles in the 2- to 5-ton range and are used for materials transport associated with the project that is not handled by vendors. The bed of the truck is flat and enclosed with removable wooden side panels.

SUMMARY

Knowledge of the types and capabilities of equipment used in construction is essential for an estimator. For many tasks there are several combinations of labor and equipment possible. It is up to the estimator to select the most economical combination and properly incorporate its cost into the estimate. In this chapter, the items of equipment most often found on a construction site were described. The listing is by no means all-inclusive. For specialty work such as paving, railroad construction, or tunneling, complete families of special-purpose equipment will be used.

REVIEW

Exercises

1. You are designated construction manager for construction of the earthwork portion of a section of new highway. Your contract includes the following:
   a. Clearing and grubbing.
   b. Subgrade shaping (which involves considerable cut and fill of a clayey-gravel).
   c. Excavation, hauling, and placement of a base layer of select material obtained from a nearby hillside borrow site.
   d. Hauling and placement of an upper base course of crushed rock obtained from a nearby commercial gravel plant.
   e. Final shaping of the road cross-section prior to paving.
   Develop a list of equipment that you will need for this project.
2. As above, except the project is a reinforced concrete overpass over the new highway and the contract is for the structure only. Concrete will be delivered from a commercial batch plant.
3. As in Exercise 1, except the project is an office building on which you are the general contractor but will subcontract specialty work.

Questions in Review

1. Why are there so many styles of compaction equipment?
2. What equipment would you recommend for the unloading and transport of gravel that is delivered in railroad cars?
3. Which is better for stockpile work: a front loader or a clamshell? Why?
4. On a fill area with relatively steep side slopes, what equipment would you use to dress up the banks before sodding?
5. You have a contract for channel improvement of a section of a river through a community. What equipment would you use for removal of debris and the shaping of this channel?

Discussion Questions

1. In terms of equipment requirements, why is it advantageous for a general contractor to subcontract specialty work?
2. A contractor has frequent but not continuous need for a small crane, a $1\frac{1}{2}$-CY backhoe, and a piece of equipment of about $\frac{3}{4}$-CY capacity for loading gravel into trucks from a stockpile. He is considering the following options:
   a. Buy a single crane unit of $1\frac{1}{2}$-CY capacity with crane, backhoe, and clamshell attachments.
   b. Buy a $1\frac{1}{2}$-CY excavator, a hydraulic crane, and a $\frac{3}{4}$-CY front loader.
   Discuss the factors that must be considered in making an economic choice between these two options.
3. A small truck-mounted crane and a hydraulic crane of the type shown in Fig. 9-8 are both excellent for handling short-term lifting requirements. Why would a contractor choose one over the other?

CONSTRUCTION
EQUIPMENT COSTS

10

INTRODUCTION

The construction equipment used on a project may be owned, leased, or rented by a contractor. On some projects, the client chooses to buy and furnish the equipment to be used by the contractor. On a project with many subcontractors, a general contractor may choose to provide the cranes and other items required regularly by a number of contractors to avoid cluttering the project site with identical types of equipment which would see low individual usage and interfere with each other. The costs of contractor-provided equipment to a project must be passed on to the client in the form of equipment charges that are included within a bid or among allowable charges on a cost-plus type of contract. This chapter will describe procedures for developing appropriate charges.

CONTRACTOR-OWNED EQUIPMENT

Contractor-owned equipment includes all items of construction equipment that have been purchased by a contractor, either outright or through installment purchase, and which are carried as assets on company accounting records. This equipment is registered in the contractor's name and the contractor may use, dispose of, or rent to another contractor any of the items at any time.

An owner of construction equipment should make the decision to buy equipment, as opposed to leasing or renting it, based upon an economic analysis of the alternatives. Included in this analysis would be such considerations as availability of cash for down payment or outright purchase, current interest rates on loans or installment purchase plans, current rental and lease rates, income tax benefits, expected usage of the equipment, and the expected life of the equipment. Ownership of equipment can be an economic drain if the equipment is underutilized, becomes obsolescent, or is subject to frequent breakdown as so often happens when kept too long.

Categories of Cost

There are two broad categories of cost associated with equipment: ownership costs and operating costs. Ownership costs also are called fixed costs or product costs while operating costs also are called variable or period costs. The ownership costs are those costs which will accrue to the owner whether the equipment operates or not. Included in this category are

1. Depreciation
2. Interest charges on borrowed money or installment purchase
3. Taxes
4. Permits and licenses
5. Storage/parking facilities
6. Insurance

Operating costs are those costs directly related to the use of the equipment and include

1. Maintenance direct costs: labor, parts, assemblies
2. Operating expendables: fuel, oil, lubricants, tires, cable
3. Maintenance overhead: buildings and yards, maintenance supervisors, maintenance equipment, utilities

All costs, both ownership and operating, must be recovered through charges for equipment use on the various company projects. These charges are generally allocated as a cost per period of use—hour, day, or month.

Operator costs are not normally included in equipment use charge rate calculations; these are included with other labor costs. Also, these rates should not include any profit or general overhead markup.

CHARGE RATE DETERMINATIONS
FOR OWNED EQUIPMENT

Cost Variation Over the Life of Equipment

Before discussing the procedures for handling the ownership and operating costs, it is useful to diagram the nature of these costs as they generally occur over the life of a piece of equipment. Using an example, assume that the original purchase price is $100,000, salvage value is 10% after a life of 5 years and 6000 hours of use, maintenance costs equal 80% of the original purchase price over the life of equipment, and all other costs including fuel and oil, amount to 135% of the purchase price over the life of item. These assumptions are reasonably representative for a major item of equipment. Figure 10-1 illustrates the example assuming that there is zero inflation during the equipment's life. Maintenance costs, represented by Curve A, will increase with age of the equipment so the cumulative cost curve is exponential in form. Depreciation (based on resale value only) is high in the first year, less in the second, and continues at an ever-decreasing rate until salvage value is reached as shown by Curve B. Other costs are relatively constant with the no-inflation assumption as indicated by Curve C. The total cumulative costs are plotted along Curve D. With a zero inflation situation, the charge rate determination procedure is simple

$$\text{Charge rate per unit of time} = \frac{\text{Total estimated costs}}{\text{Total estimated units of use}}$$

$$= \frac{\$305,000}{6000 \text{ hours}}$$

$$= \$50.83/\text{hour}$$

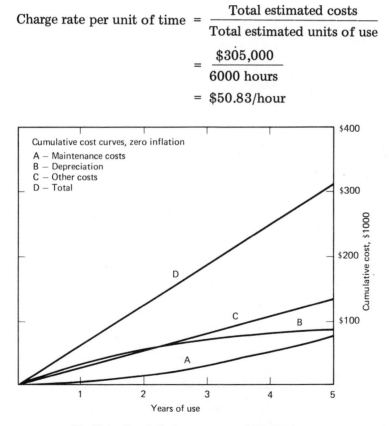

Fig. 10-1   Zero inflation cost curves, $100,000 item

All estimated costs are based upon past experience with similar items of equipment. The estimated annual use (hours, days, or other unit) is also based on historical experience.

Since zero inflation over the life of a piece of equipment is highly unlikely, the effects of inflation must be included in any cost system. Figure 10-2 takes the same piece of equipment, assumes that the maintenance and replacement costs inflate at an annual rate of 13% and all other costs as a group inflate at 9%. In this situation, the impact of shrinking dollar value over the life of the item yields a total cumulative cost of about $400,000 as compared to $305,000 under zero inflation assumptions. If a contractor had the wisdom to determine future inflation he or she could introduce appropriate inflation factors, determine total cumulative costs, and apply the straight-line formula given earlier for zero inflation to determine a charge rate. However, this is not realistic. Dollars in the first year of life of the equipment are more valuable than inflated dollars of following years. Clients would be overcharged in those early years and undercharged in later years. The reasonable so-

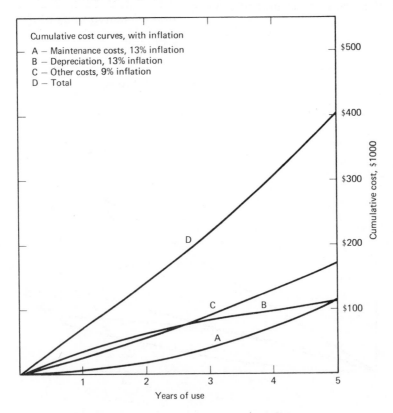

Fig. 10-2   Cost curves with inflation, $100,000 item.

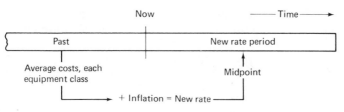

Fig. 10-3   Rate development approach.

lution is to break the life of the equipment into manageable periods and to recalculate charge rates at the beginning of each new period. Rates should be recalculated at least annually; in times of severe inflation, a contractor should consider quarterly or semiannual revisions. Once the period is chosen, the estimator projects all costs to the midpoint of the new rating period using appropriate inflationary factors as illustrated in Fig. 10-3. The procedure for handling each of the ownership and operating components is discussed in following paragraphs. For reasons which will become apparent later, some costs are computed on an annual basis, others on an hourly basis. Also in this discussion it is assumed that rates are being developed for the midpoint of the coming year. Minor adjustments permit development of rates for periods shorter than a year.

Annual Cost Group

Depreciation.   In Chap. 3, depreciation accounting for income tax and financial record keeping was discussed. In that discussion it was stated that an item could not be depreciated beyond its purchase price less salvage value (if it was, the additional amount becomes taxable income). This requirement does not hold true for calculation of the depreciation component of ownership cost. In fact, if that philosophy is followed, a contractor is undercharging. The philosophy to follow is very simple: determine what 1 year of life of an item of equipment is worth expressed in dollars of the rate year. To do this the estimator first determines what the anticipated replacement cost of the item of equipment will be at the midpoint of the rate period. This information may be obtainable from a dealer. If future prices have yet to be announced the estimator can use the most recent price figure and inflate it to the midpoint using the expected inflation rate for that type of equipment. From this future price the estimator should deduct an expected salvage value. The remainder is then divided by the economic life (number of years the equipment will be used). This yields the value of 1 year of life of the item expressed in dollars of the rate year. For example, assume a category of equipment is inflating at an annual rate of 12%, a certain model of this category cost $80,000 new at the midpoint of the past year, and typical salvage values after 5 years of operation are 10% of replacement cost. The depreciation component of ownership cost to be used for next year's rate is calculated as

$$\text{Depreciation} = \frac{(1.12) \times (\$80,000 - \$8000)}{5 \text{ years}} = \$16,128 \text{ for the year}$$

Interest Charges. Equipment purchased with loaned funds or on an installment purchase plan will have interest charges as an item of ownership expense. In any loan or purchase agreement, these interest charges will be defined in amount. A simple procedure for handling interest charges is to average them over the economic life of the equipment. The following formula applies:

$$\text{Interest charge per year} = \frac{\text{Total interest to be paid}}{\text{Economic life in years}}$$

An alternate approach yielding approximately the same charge utilizes the average unpaid principal (AUP) and then converts this to an annual or hourly interest charge as follows:

$$\text{AUP} = \frac{n+1}{2n} \times p$$

where: p = purchase price less down payment
n = economic life in years

$$\text{Interest charge per year} = (\text{AUP}) \times i$$

where: i = interest rate on loan.

These approaches are reasonably accurate as long as the installment purchase or loan payback period is approximately the same as the economic life. If the payback period is less, the calculated values are low. For example, a contractor buying a piece of equipment on a 3-year installment purchase plan and who averages interest over a 5-year economic life would be justified in increasing the interest charges to account for the difference due to time-value of the dollars paid out during the first 3 years but not charged against a project until the last 2 years.

Some cost personnel agree that interest charges should be included on all owned equipment as well as on financed equipment since capital is being tied up which could otherwise have been invested elsewhere to yield a return. It is true that a contractor should receive a fair rate-of-return on all investments. However, as explained in Chap. 3, the profit portion of the markup that is applied to direct costs in the finalization of a bid is designed to provide this return on investment. To also include an interest factor on owned equipment in a charge rate structure is equivalent to adding profit (return) twice during the estimating process since the value of this ownership is incorporated within the amount of owner's equity on the balance sheet.

Not all contractors maintain a high value of owner's equity with respect to contract volume; many building contractors are in this group. They are highly leveraged because they subcontract most of the work. These contractors may prefer to incorporate interest charges on all owned equipment in their charge rate structure and consider an additional overall profit markup on the total contract as a separate action. Heavy/ highway and industrial contractors will tend to have significant owner's equity and this is a logical base for their profit calculations.

On some contracts, a contractor is not permitted to include interest in equipment charges. However, interest charges are still a cost of doing business and must be covered in some fashion. They can be included in general overhead or reflected in the markup.

Taxes, Permits, Licenses, and Insurance.   All of these costs are either exactly known or relatively easy to project for the coming rate year.

Storage or Parking Facilities.   An owner of equipment must provide space for storage or parking of the items during periods when not on a project site. To provide such space, the owner must purchase land and buildings. This is a business expense that must be accounted for in developing construction bids. Contractors have at least two choices in this matter. First, they can choose to include the costs associated with storage and parking in the general overhead (main office) expenses that are distributed to the various projects. Second, they can attempt to isolate them and include them in the individual equipment use charges. Whatever the choice, the following approaches on the owned components are suggested:

1. *Land for Parking.*   Land does not depreciate for accounting purposes. While it may actually appreciate, the assumption is made that its value remains the same as original purchase price. As was discussed in Chap. 3, the owned portion of any land will be reflected in the total value of owner equity or investment in the firm and it is this equity or investment that is a basis for profit markup. Thus, the cost of using owned land for parking of construction equipment is accounted for in the profit markup of bids and need not be otherwise considered. However, if the land is still being purchased through a mortgage loan, the interest portion of that loan will be an expense that must be allocated in some fashion.

If land is used for parking of many pieces of equipment plus maintenance and office space, the easiest solution is to include the interest costs in the general overhead. Otherwise, a complicated allocation process must be used to assign a portion of the interest cost to each item of equipment and function.

2. *Buildings.* Unlike land, buildings are depreciated in financial accounting and for tax purposes despite the fact that buildings may not depreciate at all or increase in value. Thus, building depreciation may or may not be included as a component of equipment cost if those buildings serve the equipment population. If included, it is most easily handled as an overhead cost rather than attempting to allocate it to each item in the equipment fleet. If the building is being purchased with a mortgage loan, the interest portion is handled as for land.

Maintenance Direct Costs. Included in this group are the mechanics employee-hours, parts, replacement assemblies, maintenance expendables (such as solvents and welding supplies), and any contract maintenance work. It does not include operating expendables such as fuel. These maintenance costs will tend to increase as the equipment gets older but are basically unpredictable over short periods. With experience a contractor will learn to evaluate lifetime maintenance costs in terms of a multiplier of the current replacement cost. One procedure, yet to be discussed, provides experience multipliers in the range of 10% to 27% of current replacement cost per year. These are for average conditions and must be adjusted to reflect manufacturer, severity of use, and company maintenance procedures. Company experience is the best guide on maintenance costs and they are best handled as an average over the life of the equipment. For example, assume that past experience has shown that maintenance costs average 20% of current replacement cost per year over the life of an item. In developing the maintenance component of the rate structure for the coming year, the estimator takes the projected replacement cost of the equipment at the midpoint of the rate year (see earlier discussion of Depreciation) and multiplies it by 20%.

Maintenance Overhead. Within this category are maintenance supervisory personnel, shops, equipment used in maintenance (air compressors, lubricators, welding machines) and all small maintenance tools. Annual overhead costs include salaries plus payroll burden for included personnel, depreciation on vehicles and buildings, building utilities and fuel, shop expendables, and a charge for tool use and loss. Some contractors do not maintain a force account maintenance capability so this category is not applicable as such. Major contractors with a full equipment department maintenance facility will experience maintenance overhead costs that are 30% to 40% of maintenance direct costs.

Hourly Cost Group

Fuel. Fuel consumption by an engine is a function of its brake horsepower (bhp), load factor (LF), and state of maintenance (maintenance factor, MF). The basic equation is

$$\text{Fuel consumption/hour} = (\text{bhp})(\text{gal/bhp/hour})(\text{LF})(\text{MF})$$

Gasoline engines at full load will burn about 0.1 gal/bhp/hour and diesel engines about 0.06 gal/bhp/hour. However, none operate a full hour at full load so a load factor is applied. Some representative load factors are

Wheel mounted equipment, highway	25–40%
Wheel mounted equipment, off-highway	50–60%
Track dozers and loaders	50–75%
Cranes	30–50%
Clamshells and draglines	40–60%
Shovels and excavators	50–70%

Adjustment for state of maintenance is up to the contractor. With no adjustment, this factor is 1.0; old or poorly maintained equipment is represented by maintenance factors in excess of 1.0. Fuel consumption is converted to a cost/hour by multiplying by the anticipated cost of fuel per gallon at the midpoint of the rate period.

Lubricants, Filters, Hydraulic Oil. Under average conditions lubricating oil is changed every 100 to 125 hours unless otherwise recommended by a manufacturer. Changes will be more frequent if the operating atmosphere has been particularly dusty. Oil filters are changed with the oil; air and other filters changed as recommended by the manufacturer. Hydraulic oil requirements are relatively minor. For any items changed on the basis of hours of use the applicable formula for determining hourly cost is

$$\text{Hourly cost} = \frac{\text{Replacement cost}}{\text{Hours between replacement}}$$

An alternate approach is to relate the cost of these items to the basic fuel cost. A common rule-of-thumb is to increase the fuel cost by one-third for gasoline engines and one-half for diesel engines to account for the cost of this group. As with all such rules-of thumb, considerable variation in practice can be expected. Reference to manufacturer's handbooks may provide more exact estimating guides.

Tires, Crusher Jaws, and Cutting Edges. Such components wear out in proportion to hours of service and are routinely replaced when worn. A typical off-highway tire will last 2000 to 5000 hours depending upon equipment type and severity of use. Similar ranges will be experienced for other replaceable components. The hourly cost to use for these is found from the formula

$$\text{Hourly cost, replaceable components} = \frac{\text{Replacement cost}}{\text{Expected life in hours}}$$

RECOMMENDED CHARGE RATE STRUCTURE

In the previous section, some elements of cost were expressed as annual figures, others as hourly figures. This was done to lead into a recommended system for charging jobs for equipment. The items included in the annual group were all ownership components of cost plus major maintenance and maintenance overhead from the operating group. The hourly group contained only those items most directly related to hours of use.

The recommended system for charging equipment is very similar to that used by utilities and phone companies: The rate structure consists of an availability charge and a use charge. The availability charge, as its name implies, is for making a piece of equipment available to a job for a period of time, whether it is actually used or not. The use charge is for actual operation of the piece of equipment. This approach recognizes that it is impractical to move many pieces of construction equipment from job to job every day or so. For example, a fleet of earthmoving equipment assigned to a road project will remain there until no longer needed even though there will be many nonworking days on that job. Transfer of the equipment to another job during these nonworking periods is normally not practical because of distance, lack of adequate transport vehicles, or cost. Still, every day that these items of equipment are on the contractor's inventory, they are costing money and this money must be recovered from projects. The logical solution is to charge the client an availability charge for all time that the equipment is committed to a job. This availability charge includes all cost components identified in the annual cost group and will be a daily, weekly, or monthly charge. To arrive at the availability charge, all cost components in the annual group are added together to get a total annual cost. Then, the estimator must make an assumption of the number of days, weeks, or months that the equipment is normally assigned to jobs each year. This will always be less than a full year since a contractor can never count on 100% equipment utilization due to type of equipment, weather, maintenance downtime, or lack of work. Equipment such as asphalt paving machines, power brooms, or other very specialized equipment may be used only 4 or 5 months a year. Equipment such as office trailers and utility trucks may find almost full year use. Having made an appropriate assumption as to the expected annual commitment, the estimator divides the total annual cost by this number to arrive at an availability cost per unit of time.

The use charge is an hourly charge and is the sum of all items discussed under the hourly charge group. It will be charged for every hour of actual or estimated equipment operation.

The two charges are put to use in an estimate in this way. First, the estimator obtains from the intended construction manager, or develops on his or her own, a list of equipment requirements for the job. These are scheduled against time over the life of the job. Figure 10-4 is a simplified version of such a listing. Then, based on this schedule, the number of charge periods for each item are determined and multiplied by the availability rate to get the total availability charge for the project estimate. Next, the estimator makes an estimate of the number of actual hours of use of the equipment during the committed period. This quantity, when multiplied by the hourly use charge, gives the second component of cost for the estimate. When totalled, the result is the cost estimate for that item. Again, note Fig. 10-4 to see how this is accomplished.

ALTERNATE CHARGE RATE PROCEDURES

Procedures other than that just described are available to the contractor. They do not necessarily yield the same cost figures as the recommended procedure so a contractor must carefully evaluate all possibilities to select the one which best serves his or her need and most accurately spreads true costs over the jobs.

Minnesota AGC Worksheet

The worksheet shown in Fig. 10-5 was extracted from the manual, *Construction Equipment Cost Manual for Value Recovery Rates*, prepared by the Associated General Contractors of Minnesota, 111 East Kellogg Boulevard, Saint Paul, Minnesota 55101. The data it contains were derived from member experience with common types of construction equipment. An excellent feature of this manual is that all costs are related to the current replacement cost of a piece of equipment; thus, inflation is automatically considered although an inherent assumption is that all costs inflate at the same rate as replacement costs. This manual is intended to be used only as a guide since cost factors presented represent average conditions and considerable variation can be expected from

Item	Monthly avail. $	Hourly cost $	Scheduled commitment												Months avail.	Hours use	Total estimated cost
			J	F	M	A	M	J	J	A	S	O	N	D			
Pickup, $\frac{3}{4}$ T, 4X4	$300	$3.00													6	360	$2880
Office trailer, 20 ft	$100	NA													6	NA	$600
Crane, hyd, 5 ton	$1200	$4.00													3	300	$4800
Loader, wheel, $\frac{3}{4}$ CY	$650	$3.25													2	200	$1950

Fig. 10-4   Equipment schedule and cost estimate.

174

contractor to contractor. It is particularly valuable for use by design engineers in preparing an engineer's estimate of cost since they must rely on handbook data for their estimates. The cost categories included in the Minnesota AGC system incorporate all those already discussed and, as before, there is no markup for profit or general overhead. The various columns of the rate schedule extract shown in Fig. 10-5 have the following meanings and use:

Column 1 *Replacement Value Recovery Rate.* This entry accounts for depreciation as discussed in this chapter, not that for income tax or financial records. A multiplier of 20% implies a 5-year life; a 33% multiplier is for a 3-year life. These multipliers are for average conditions and use as reflected in columns 9 and 13. Should actual conditions differ significantly, the user may consider proportional adjustments.

Column 2 *Interest, Insurance, Storage.* These ownership costs are grouped and assigned a combined multiplier of 12% in most cases which includes an 8.5% interest rate, 1% insurance rate, and 2.5% storage rate. If actual rates differ significantly, appropriate adjustments can be made. In the case of interest on equipment already completely owned, the contractor may choose to delete the interest component entirely as suggested in earlier discussion.

Column 3 *Taxes and Licenses.* This space is left blank because rates vary widely. The user must insert an appropriate percentage.

Column 4 *Total Fixed Cost.* This column is the sum of columns 1, 2, and 3.

Column 5 *Major Components and Repairs.* Direct costs associated with major maintenance are accounted for here including mechanics, parts, tools, and shop expendables. Tires are included in this group.

Column 6 *Operating Expendables.* This factor accounts for fuel, oil, filters, cutting edges, cables, batteries, and similar expendables plus labor for installation.

Column 7 *Supporting Facilities.* Maintenance and operating overhead are accounted for here. Included are storage yards, shops, maintenance vehicles, equipment transportation, and the cost of maintaining parts inventories.

Column 8 *Total Variable Costs.* This is the sum of columns 5, 6, and 7.

Column 9 *Average Earning Months per Year.* This is the assumed number of months that the item is committed to projects each year. Used in conjunction with column 13, the assumed annual hours of use may be determined.

Column 10 *New Equipment Replacement Value.* In this column is entered the cost of the item at the present time or midpoint of rating

VALUE RECOVERY RATE COST SCHEDULE

EQUIPMENT DESCRIPTION GROUPING BY CLASS	VALUE RECOVERY FOR ANNUAL FIXED COSTS AVERAGE ANNUAL COST Percent of New Replacement Value			
	New Replacement Value Recovery Rate **1**	Interest, Insurance, Storage **2**	Freight, Taxes, Licenses, and Sales Tax **3**	Total Fixed Cost* **4** *(1+2+3)=4
2.15 SCRAPERS (Earthmovers)				
1) Towed	20%	+ 12%	+_____%	=_____%
2) Self-Propelled	20%	+ 12%	+_____%	=_____%
2.16 TRACTORS (Including accessories such as dozer blades and other attachments)				
1) Crawler	20%	+ 12%	+_____%	=_____%
2) Rubber Tired	20%	+ 12%	+_____%	=_____%
2.17 TRAILERS				
1) Hauling (Bottom, side and rear dump, water wagons, high-boys and low-boys)	20%	+ 12%	+_____%	=_____%
2) Portable field offices and storage vans	20%	+ 12%	+_____%	=_____%
2.18 TRUCKS (Tractors, Dump Trucks and Utility)				
1) On highway (low axle weights)	25%	+ 12%	+_____%	=_____%
2) Off Highway (high axle weights)	20%	+ 12%	+_____%	=_____%
3) General purpose utility trucks (Pickups etc.)	25%	+ 12%	+_____%	=_____%

Fig. 10-5 Sample Minnesota AGC equipment rate schedule. (Courtesy AGC of Minnesota)

VALUE RECOVERY FOR ANNUAL MAINTENANCE & SUPPORTING FACILITIES COSTS
AVERAGE ANNUAL COST
Percent of New Replacement Value

APPLICATION OF COST SCHEDULE
Dollar Values

Major Components & Repairs Recovery	Operating Expendables	Supporting Facilities	Total Maintaining & Supporting Costs*	Average Number Earning Months Per Year	New Equipment Replacement Value*	Cost Per Earning Month*	Hourly Use Rate*	Average Hours Each Earnings Month
5	**6**	**7**	**8**	**9**	**10**	**11**	**12**	**13**
			*(5+6+7)=8		F.O.B. Factory	$\frac{*(4+8)\times10}{9}=11$	*11÷13=12	
7%	+ 2%	+ 3%	= 12%	7	$_____	$_____	$_____	160
11%	+ 6%	+ 3%	= 20%	8	$_____	$_____	$_____	160
11%	+ 9%	+ 3%	= 23%	8	$_____	$_____	$_____	160
10%	+ 7%	+ 3%	= 20%	8	$_____	$_____	$_____	160
14%	+ 4%	+ 3%	= 21%	7	$_____	$_____	$_____	160
5%	+ 3%	+ 3%	= 11%	10	$_____	$_____	$_____	160
13%	+ 9%	+ 3%	= 25%	6	$_____	$_____	$_____	160
8%	+ 6%	+ 3%	= 17%	8	$_____	$_____	$_____	160
13%	+ 5%	+ 3%	= 21%	7	$_____	$_____	$_____	130

Fig. 10-5 *Continued*

period. It should include full delivery price including taxes, but less tires and licenses.

Column 11   *Cost per Earning Month.* The replacement price (column 10) is multiplied by the sum of the percentage multipliers derived in columns 4 and 8. This quantity is then divided by earning months (column 9) to give the cost per earning month.

Column 12   *Hourly Use Rate.* Column 11 when divided by the earning hours per month (column 13) gives an hourly charge rate.

Column 13   *Average Hours Each Earnings Month.* This column provides the assumed number of earning hours per earning month for this type of equipment.

The ultimate products of this method are costs per earning month and costs per earning hour. These products are not the same as the monthly availability and hourly use charges discussed earlier in the chapter. However, the worksheet can be adapted to develop these charges as

Availability charge/month

$$= \frac{(\text{Column 10}) \, (\text{Column 4} \, + \, \text{Column 5} \, + \, \text{Column 7})}{\text{Column 9}}$$

Hourly Charge

$$= \frac{(\text{Column 6}) \, (\text{Column 10})}{(\text{Column 9}) \, (\text{Column 13})}$$

This approach accounts for tires as part of the availability charge rather than as an hourly charge. Otherwise cost components are arranged similarly.

Equipment Manufacturers and Association Handbooks

Worksheets for the development of charge rates for equipment are available from some manufacturers and equipment associations. Figure 10-6 is a worksheet contained in Power Crane and Shovel Association handbook, *Operating Cost Guide for Estimating Costs of Owning and Operating Cranes and Excavators.* This worksheet includes items already discussed including provisions for an inflation factor (Item No. 4). It also has a section for determining the crew costs.

Typical of worksheets available from manufacturers is that shown in Fig. 10-7 which was taken from the *Caterpillar Performance Hand-*

OPERATING COST ESTIMATING FORM

Published by the Power Crane and Shovel Association
For Use with PCSA Operating Cost Guide

Machine: _____

Cost F.O.B. Factory ....$ _____

Additional Equipment ....$ _____

Shipping or Transporting Cost ....$ _____

Unloading, Erecting & Moving to Site ....$ _____

Item No. 1 — Total Cost or Investment ....$ _____

Item No. 2 — Salvage Value ....$ _____

Item No. 3 — Depreciable Value (Item 1 minus Item 2) ....$ _____

Item No. 4 — Inflation or Deflation Factor (see page 11)
      cumulative annually — ± _____ % of Item No. 3 ....$ _____

Item No. 5 — Adjusted Depreciable Value ....$ _____

Item No. 6 — Economic or Useful Life: _____ years, or _____ hours.

Item No. 7 — Average Investment (see formula, page 13): $ _____

	Cost	
	Per Year	Per Hour
Item No. 8 — Depreciation: _____ % of Item No. 5 (Table 2)..................$ _____		$ _____
Item No. 9 — Interest, Taxes, Insurance, Storage: _____ % of Item No. 7.......$ _____		$ _____
Item No. 10 — Repairs, Maintenance & Supplies: _____ % of Item No. 8 (Table No. 3) ....$ _____		$ _____
Item No. 11 — Total Fixed Costs (Items 8, 9, 10) ....$ _____		$ _____

Item No. 12 — Engine Fuel & Lubricating Oil:

Fuel per Hour .... _____ Gal. @ _____ = $ _____

Lubricating Oil per Hour _____ Gal. @ _____ = $ _____

Auxiliary Fuel & Lubricating Oil per Hour ....$ _____

| Total Fuel and Lubricating Oil ....$ _____ | | $ _____ |

Item No. 13 — Direct Labor — Operating Crew

_____ Operators @ $_____ per hour = $ _____

_____ Oiler @ $_____ per hour = $ _____

_____ Other @ $_____ per hour = $ _____

| Total Direct Labor Costs ....$ _____ | | $ _____ |

Item No. 14 — Total Direct Costs (Items 11, 12, 13) ....$ _____		$ _____
Item No. 15 — Indirect Costs: Supervision, Overhead, Profit, etc. ....$ _____		$ _____
Item No. 16 — TOTAL COSTS (Item 14 & 15) ....$ _____		$ _____

Fig. 10-6   PCSA cost estimating form. (Courtesy PCSA)

Machine Designation .... _____ _____ _____

DEPRECIATION VALUE

  1. Delivered Price (including attachments) ... _____ _____ _____

  2. Less Tire Replacement Costs:

      Front _____ _____ _____

      Drive _____ _____ _____

      Rear _____ _____ _____ ... _____ _____ _____

  3. Delivered Price Less Tires .... _____ _____ _____

  4. (Optional) Less Resale Value or Trade-in _____ _____ _____

  5. Net Value for Depreciation .... _____ _____ _____

OWNING COSTS

  6. Depreciation: Net Depreciation Value (Item 5)

               Depreciation Period in Hours

$$\frac{\text{Value}}{\text{Hours}}$$ _____ _____ _____ ... _____ _____ _____

  7. Interest, Insurance, Taxes:

      Annual Rates: Int.__% Ins.__% Taxes__%

      Estimated Annual Use in Hours _____

$$\frac{\text{Factor x Delivered Price (Item 1)}}{1000}$$

$$\frac{\text{x}}{1000} \quad \frac{\text{x}}{1000} \quad \frac{\text{x}}{1000}$$ _____ _____ _____

  8. TOTAL HOURLY OWNING COSTS ..... _____ _____ _____

OPERATING COSTS

      Unit Price x Consumption

  9. Fuel: _____ x _____ _____ _____

  10. Lubricants, Filters, Grease:

      Unit Price x Consumption

    Engine _____ x _____ _____ _____

    Transmission _____ x _____ _____ _____

    Final Drives _____ x _____ _____ _____

    Hydraulics _____ x _____ _____ _____

    Grease _____ x _____ _____ _____

    Filters _____ x _____ _____ _____

                    Sub Total

  11. Tires: $$\frac{\text{Replacement Cost}}{\text{Estimated Life in Hours}}$$

$$\frac{\text{Cost}}{\text{Life}}$$: _____ _____ _____ .... _____ _____ _____

  12. Repairs: $$\frac{\text{Factor x Del. Price Less Tires}}{1000}$$

$$\frac{\text{x}}{1000} \quad \frac{\text{x}}{1000} \quad \frac{\text{x}}{1000}$$ _____ _____ _____

  13. Special Items:

_____ _____ _____

  14. TOTAL HOURLY OPERATING COSTS.... _____ _____ _____

  15. Operator's Hourly Wage _____ _____ _____

  16. TOTAL HOURLY OWNING AND

      OPERATING COSTS .... _____ _____ _____

Fig. 10-7 Caterpillar Tractor Co. estimating worksheet. (Courtesy Caterpillar Tractor Co.)

book, published by the Caterpillar Tractor Co. This worksheet does not specifically include any consideration of inflation but inflation can be applied by use of replacement costs instead of delivered prices. This particular worksheet yields an overall cost per hour. An entry is included for operator wages if desired. Procedures for completing the form are detailed in the *Handbook*.

SUPPORTING EQUIPMENT COST-ESTIMATING INFORMATION

Estimating Salvage Value

Small pieces of motorized equipment such as chain saws and small tractors are generally in such poor condition at the end of their economic lives that a zero salvage value can be assumed. For major items of equipment, the resale value can equal or exceed original purchase price in times of severe inflation. Contractors will probably find it best to relate salvage value to current replacement cost, as was done in an earlier example, rather than to original purchase price.

Projecting Future Equipment Costs

The Bureau of Labor Statistics (BLS), US Department of Labor, surveys the market and prepares and makes available periodic statistics on the cost of construction equipment. *Engineering News-Record* magazine also publishes the BLS statistics in their quarterly cost roundup issues. The statistics given include a composite cost index (overall average) as well as individual cost indices for many categories of equipment. The base for the index system is 1967 which is assigned an index of 100.0. Following are the composite equipment cost indices from 1967 through 1979 (January of each year).

Year	Index	% Change from Previous Year	Year	Index	% Change from Previous Year
1967	100.0	—	1974	152.3	+16.5
1968	105.7	+5.7	1975	177.2	+16.4
1969	110.4	+4.5	1976	193.3	+ 9.1
1970	115.9	+5.0	1977	208.8	+ 8.0
1971	121.4	+4.8	1978	222.6	+ 6.6
1972	125.7	+3.5	1979	245.2	+10.2
1973	130.7	+4.0			

Review of these indices shows that prices increased at a low rate between 1967 and 1973 (average growth rate of 4.5% annually) while they increased sharply thereafter (averaging 11% annual increase to 1979). An estimator must project inflation (or deflation) based on his or her best judgment or the recommendation of professional procurement specialists.

Determining Annual Operating Hours

A key assumption in all estimating methods is the number of hours, weeks, or months of operation of a piece of equipment during the year. The maximum number of hours is 2080 (52 weeks/year × 40 working hours/week) for single-shift operation. For multiple shifts it would be proportionally greater. However, lack of work, weather, holidays, downtime, nature of the equipment and other factors reduce these figures. Holiday periods are easily identified. Lacking past experience in an area, an estimator can research climatological records to estimate downtime due to weather for a given type of equipment or consult with potential subcontractors in the area. Maintenance downtime is a function of job operating conditions and the effectiveness of the company's maintenance program but a downtime of 10% is a reasonable starting assumption for major items of equipment and 20% for small motorized items. Overall, the maximum number of hours that an item is available and able to perform work (assuming a 40-hour week) would range from about 1300 hours in those areas of the United States with extremes of cold and precipitation to about 1800 hours in the warmer, drier areas. From these figures you must subtract days lost for lack of work, scheduling problems, late material deliveries, operator absence, waiting for prerequisite work to be completed, etc. In the Minnesota AGC method, a common assumption for annual working hours is 1120 (7 months × 160 hours per month). Specialty equipment such as concrete batch plants, pile drivers, power brooms, and tunneling machines have significantly lower utility.

Estimating Economic Life

The term economic life has been used often in previous discussions and it has been assumed to equal the number of years that a piece of equipment is kept in use before disposal. In theory, the economic life of equipment is that lifespan during which incremental income either exceeds or equals incremental costs. In plain language, you dispose of an item when it costs more to keep than it earns. Various systems have been devised to calculate economic life but each is subject to a number of assumptions regarding costs, rate of use, charge rate, profit markup, and similar considerations. In practice, there is no exact procedure for determining economic life. If there is any numerical criterion to be applied to economic life, it should be that the item of equipment is available for operation 90% of the time. In other words, if maintenance downtime reaches the point that the contractor cannot count on its availability 90% of the time, it should be disposed of. For many major items of construction equipment, an economic life of 4 to 5 years is common. In the case of large cranes, economic lives of 10 or more years are not uncommon. Under conditions of severe usage, a major piece of equipment may exhaust its economic life in as few as 2 years. For power hand tools, the life is probably a year; for installed shop tools, many years. The Minnesota AGC worksheets (Fig. 10-5) provide in column 1 an indication of contractor evaluation of economic life for each equipment type.

Rebuilding Equipment

Rapidly rising costs of new construction equipment will make rebuilding an attractive option in lieu of disposal for both small and large items of equipment. The costs of rebuilding are equivalent to a new capital investment which a contractor can depreciate for tax purposes. For charge rate purposes, the contractor will treat the rebuild costs as a purchase price and extend the economic life a reasonable number of years—usually less than that for new equipment. Maintenance costs should be adjusted upward since the rebuilt item cannot be expected to be as reliable as a new one.

Equipment Fleets

In Chap. 9 it was mentioned that larger contractors normally have an equipment department which owns, maintains, and rents equipment to the jobs. Such departments go through the procedures discussed earlier to determine charge rates and publish periodic catalogs of rates for use by estimators and construction managers. Such equipment departments have the advantage of owning many pieces of equipment of each class, where a class represents a given size and type of equipment without regard to manufacturer or age. An equipment department will try to maintain a balanced distribution of equipment ages within each class so that they are not dealing with all new or all old items which will distort the maintenance picture. With such a balanced fleet, it is possible to obtain very meaningful records on utilization and costs which provide the basis for calculating charge rates. Equipment departments normally operate as nonprofit cost centers, the objective being to exactly recover all costs through rental to projects. Equipment departments have proven to be extremely cost effective. Through centralized procurement, maintenance, dispatch and record keeping and through equipment standardization of makes and models in the fleet, they are able to hold equipment charge rates to the minimum.

EQUIPMENT LEASING

Leasing is the second option for obtaining required construction equipment. Leasing is the provision of items of equipment to a contractor in return for periodic use payments. In this way it is similar to renting. However the two differ in these ways

1. A lease normally involves a long period of time, usually approximating the economic life. A rental period can be as short as a day. With a lease, the lessee is obligated to keep the equipment throughout the lease period whether it is needed or not.

2. A proposed lessee's financial strength must be about the same as that expected if he or she were to purchase equipment through a bank loan or on an installment purchase plan. This is not true of rentals.

3. There are many lease forms and payment plans that can be negotiated between the lessee and the lessor. For example, payments can be patterned to coincide with anticipated earnings. Rental plans are more rigid. Typical leases available will be discussed later.

4. In some leasing arrangements the investment tax credit is passed to the lessee. It is never passed with rentals.

5. A leasing firm does not maintain stocks of equipment. They purchase needed equipment from suppliers for delivery to the lessee.

Types of Leases

Among the more common forms of leases are

1. *True Lease.* With this lease the lessor retains title to the equipment and is considered its owner. At the expiration of the lease term, the equipment will revert to lessor control. The ITC may be passed to the lessee under this form of lease. This usually is done in exchange for correspondingly higher periodic payments.

2. *Finance Lease.* This lease gives the lessee the option to purchase the piece of equipment at some point in time. The monthly lease payments will probably be higher than in the case of a true lease, the difference reflecting purchase of equity in the item. This type of lease is essentially an installment purchase plan with a down payment requirement.

3. *Sale and Leaseback.* This form of lease finds use when a contractor is in need of working capital. The contractor sells the equipment to the leasing agency and then leases it back.

4. *Wet Lease.* This is a modified form of any lease which includes extra services such as fuel, maintenance, or operators. This form is not too common in leasing although very common with renting.

At the expiration of the lease term, a number of alternatives are possible. These are

1. The equipment is returned to the lessor.
2. A new lease on the same item is negotiated but at a lesser rate to reflect the worn condition and obsolescence of the item.
3. The contractor buys the equipment at a price guaranteed in the lease.
4. The contractor is allowed to buy the equipment at a price considered to be fair market value at the time of sale.
5. The equipment is sold on the open market by the lessor.

There are income tax and accounting implications associated with the various types of leases and a contractor should seek professional advice concerning these implications before entering into any lease agreement. Internal Revenue Service rulings have construed finance leases to be installment purchases. When so interpreted, the periodic lease payment is no longer fully deductible as a business expense and this situation must be reflected in the contractor's handling of the calculations of ownership and operating costs. This matter is beyond the scope of this text so further discussion will assume true lease conditions.

Calculation of Use Charges on Leased Equipment

Ownership Costs. With a true lease, the depreciation and interest components are replaced completely by the lease payments. Other ownership type costs still remain, namely

1. *Taxes, Licenses, Permits, and Insurance.* Certain of these may be provided by the lessor and will be defined in the lease agreement. The cost of those provided is part of the lease payment. In general, both the lessor and lessee will carry insurance on the equipment so that provided by the lessee must be treated as an added ownership cost.
2. *Parking and Storage.* These added costs are to be treated as for owned equipment.

Operating Costs. Unless the lease is wet these costs are lessee's responsibility and will be handled the same as for owned equipment.

EQUIPMENT RENTAL

Rental is the ideal option for short-term equipment needs. Items can be rented by the day, week, or month. The rates per unit of time are higher than for leasing but these rates include coverage of some costs not included within a lease and the contractor is not bound to keep the equipment any longer than required. Seldom would a rental contract extend beyond a year.

A rental agency normally will maintain an inventory of equipment so that items can be obtained on short notice. They also will have maintenance shops for handling most maintenance and repair tasks although they may handle their maintenance obligations by dispatching mobile maintenance teams or by contracting for such services.

Calculation of Use Charges on Rental Equipment

Ownership Costs. Under most rental agreements, all ownership costs are included within the rental fee. A contractor may have blanket insurance coverage which includes rental equipment so this is one possible exception. The rental agreement will specify all included goods and services.

Operating Costs. Maintenance and repair normally are rental agency responsibilities, as would be operating expendables such as tires, blade edges, etc. Fuel, oil, and lubrication costs generally will be the only operating costs (other than the operater himself) to be paid by the renter.

SMALL TOOLS

All of the discussion in this chapter has centered on the larger pieces of construction equipment. Equally essential on any job are many small tools such as hammers, hand drills, and wrenches. Calculating periodic use charges on each small tool is impractical yet some charge must be made to each job since these tools do cost money and they wear out or get lost. Some contractors require their workers to own and bring to the job their own personal tools; under some union contracts the contractor must pay a tool allowance to his or her craftsmen. In all situations, the contractor will own and furnish some of the larger powered tools. For contractor-owned tools, there are at least three ways to handle the costs.

Method 1. A value should be established for the total inventory of small tools utilized within the firm. Next, an annual percentage turnover rate for the inventory is assumed. The inventory value multiplied by the turnover rate becomes the cost per year and this cost is then included in general overhead expenses which are distributed to the various projects of the firm.

Method 2. A list of small tools is developed for each work package, such as concrete finishing, that is done by the firm and it is given a dollar value. Next, the life of this package of tools is estimated in terms of production units, such as square yards of concrete finished. The value of the tools divided by the life of the tools in terms of production units gives a cost per unit of production. This cost is then added to other costs to arrive at the total direct cost of a work package.

Method 3. Perhaps the simplest of all methods is to add a small tool cost as a percentage (about 2%) of labor cost. However, this markup should be validated over a period of time to insure that it is adequate.

RECORDING AND ACCOUNTING FOR COSTS IN AN ESTIMATE

Some of the equipment assigned to a job will be committed to specific work packages, others will support a number of work packages on a daily basis, and still others are purely administrative overhead. A contractor can choose to allocate equipment costs directly to work packages, consolidate them all and handle as a job indirect cost, or use a combination of both approaches. In general, it is much more practical to handle all administrative equipment and those items of equipment which support two or more work packages (such as cranes) as a job indirect than attempt to distribute their cost to work packages. For equipment items that are committed entirely to specific work packages (such as concrete pumps), either option is satisfactory. On unit-price estimates, all equipment committed to the production of the bid item should be fully costed within the direct work package, not as a job indirect, to insure that the unit price truly reflects all included costs.

SUMMARY

The cost of construction equipment utilized on a project is another of the direct or job indirect costs that must be accounted for. Such equipment may be owned, leased, or rented, the choice being dependent upon the expected period of equipment use and a variety of financial considerations. The costs associated with a piece of construction equipment are broadly categorized as ownership costs and operating costs. It is the estimator who must fully account for each of these categories in developing the equipment use charges to be used for a given job. There is no universally accepted method for determining these charges; as with most estimating procedures, each firm must, through trial and experience, develop a procedure which results in realistic, competitive charges. However, the following guidelines are recommended

1. Recover true costs, both cash flow and noncash flow

2. Convert all costs to dollars of the time period for which the rate is calculated

3. Use a single charge rate for all items of a given equipment class without regard to manufacturer or age of item

4. Base charges on own experience, not industry handbooks, whenever possible

5. Do not confuse depreciation accounting for tax or financial records with depreciation accounting for equipment charges

6. Consider use of a two-part charge structure. One charge is an availability charge by the day, week or month; the other is an hourly use charge

7. Change rates at least annually; more often in times of economic instability

8. Do not include operator wages or markup for profit and general overhead in the rate structure

REVIEW

Exercises

1. Data are given below for three items of equipment owned by a contractor. In each case, the equipment was purchased as installment purchase with 20% down payment and the remainder to be paid in 48 equal monthly installments at 12% interest per year. From this data, develop what you consider to be reasonable monthly availability and hourly use charges for this equipment for 1979.

| | Equipment Item | | |
	A	B	C
Date of purchase	1975	1976	1977
Purchase price (including tax and delivery)	$180,000	$10,000	$50,000
Down payment	$36,000	$2,000	$10,000
Monthly payments	$3792.21	$210.67	$1053.35
Total interest to be paid	$38,019	$2,112	$10,561
Estimated economic life	5 years	5 years	3 years
Trade-in value at end of economic life	15% of current price		
Average hours use/year	1000	1600	600
Months per year of commitment to projects	10	9	4
Maintenance as % of replacement cost/year	20	18	25
Storage/maintenance overhead	2% of current price/year		
Insurance, taxes and permits	1% of current price/year		
Fuel type	Diesel	Diesel	Diesel
Fuel consumption/hour	15 gal	4 gal	11 gal
Fuel cost	$1.00/gal		
Lubricants, filters	40% of fuel cost		
Tires: cost/change	$6000	$1500	N/A
Average life in hours	4000	2000	N/A
Expected inflation rate, 1975–1979	11% per year		

2. Using the extract from the Minnesota AGC cost handbook (Fig. 10-5), calculate the hourly use rate for a crawler tractor which has a projected replacement cost of $120,000 at the midpoint of the new rate period.

3. For the crawler tractor in Exercise 2, above, determine a monthly availability charge and an hourly use charge.

Questions in Review

1. Describe the various methods by which a contractor can obtain construction equipment.

2. What is economic life?

3. What are the two major categories of cost for equipment? What components are within each category?

4. What are the advantages of worksheets? What dangers are there in using a standard worksheet such as the Minnesota AGC worksheet?

5. Where does one find information on fuel consumption, maintenance costs, etc., for a piece of equipment?

6. What are the relative advantages and disadvantages of ownership, leasing, and rental of equipment?

Discussion Questions

1. When calculating periodic use charges for equipment shouldn't you include profit and overhead in the charge itself?

2. What can a contractor do who has owned or leased equipment that is not being used?

CONSTRUCTION
EQUIPMENT REQUIREMENTS

11

INTRODUCTION

Construction equipment is that equipment used by a contractor in the construction or servicing of a project. It is distinguished from installed or permanent equipment which is equipment that is permanently installed and becomes part of the completed project. In some cases, installed equipment is used as construction equipment. For example, overhead cranes and elevators to be incorporated in a project can be installed early to permit their use as construction equipment.

Chapter 9 described common types of construction equipment. Chapter 10 explained various ways of determining the use charges for this equipment. In this chapter we are concerned with the determination of the composition of the construction equipment fleet and its productivity.

CONSTRUCTION EQUIPMENT CLASSIFICATIONS

For purposes of this text, construction equipment is divided into three categories. The definition of the three hinge upon an understanding of the term *pay item*. A pay item is a typical unit of work used by an estimator in developing a cost estimate and by cost engineers in determining progress. Also, pay items are used as the work units on a unit-price contract. Typical pay items are cubic yards of material processed; tons of reinforcing steel placed; square feet of forms installed, etc. Each pay item's production is a function of the application of labor and/or equipment. With each pay item, certain labor or equipment is the key production force, others being support. For example, in earthwork, the power shovel, truck, or scraper is a key producer while the pushdozer, water distributor, and surveyors are support. In concrete form work, the labor assigned is the key production element with cranes, saws, and other equipment used in their erection being support. Finally, there are pieces of equipment on the job which cannot be related to any pay item. Thus, there are these three categories: production equipment, support equipment, and administrative equipment.

Production Equipment

Production equipment is that equipment whose primary function is the production of pay items. Power shovels produce cubic yards of excavation; excavators and ditchers produce linear feet of ditch; trucks move cubic yards of material; quarries produce cubic yards or tons of rock, etc. Production equipment is a direct cost item although some contractors choose to charge all equipment as a job indirect cost.

Support Equipment

Support equipment includes those items that play a support role in the production of pay items by production labor and/or production equipment; they are not primary producers. Cranes, generators, air compressors, and welding machines are generally in this category. Enough support equipment of the proper sizes must be available when needed or production labor and equipment may be kept from working or prevented from producing at capacity.

Support equipment may be a direct cost item if its operation is fully associated with a work package. If an item of support equipment provides a general support role to several work packages, it is better handled as a job indirect cost.

Administrative Equipment

Administrative equipment is associated with project administration and not directly involved with the production of pay items. Supervisor vehicles, ambulances, courier scooters, warehouse forklifts, etc., are in this category. Absence of administrative equipment items will not directly affect production on the project. Administrative equipment is charged as a job indirect cost.

SELECTING EQUIPMENT

A major task for the estimator in developing the cost estimate is the need to identify all resources required to build and administer the project. For each work package the estimator must select equipment of the right types and sizes and in the required number. In coordination with the

schedulers, these lists must be balanced to provide an economical distribution of equipment over all work packages so as to minimize numbers
of items of equipment required, to reduce idle time of this equipment,
and to meet schedules. For some work packages, the decision will be
made by the planners to subcontract the work as the more economical
approach.

Equipment Productive Time and Production Rates

Out of the total number of potential working hours in a year a
piece of equipment may actually be productive only 10 to 40% of the
time. Loss of productive time occurs in two ways. First, a piece of
equipment will be incapable of performing work a number of days each
year because it is down for maintenance, weather conditions prevent
work, jobs are not available requiring use of the item, or the equipment
is being moved. Second, when on the job and in a position to operate,
it will effectively produce less than 60 minutes each hour because of
operator breaks, refueling or minor maintenance breaks, waiting on other
operations, short moves, brief weather problems, and similar detractions.

When an item is actually producing, its rate of production is yet
another variable, it being influenced by site conditions, the nature of
materials being processed, the altitude in the case of air-breathing engines, the condition of equipment, operator ability, and comparable
items which affect speed of operation. It follows that equipment productivity, like labor productivity, presents the estimator with another major
variable which must be contended with in developing cost estimates.

Determining Productivity

On-Site. Since no two jobs are exactly alike, the ideal way of determining productivity is to measure it using company operators on
company equipment at the intended job site. Using stopwatches, measuring tapes, scales, or whatever devices may be required, the actual
performance of equipment is measured over a sufficient period to establish reasonably accurate figures. Under test conditions, the productivity
measured would normally be that for no-delay conditions and would be
adjusted to reflect breaks from work that will occur in actual practice.
Unfortunately, it is seldom possible for on-site measurements to be made
and the estimator must make an off-site determination using the best
available information plus good judgment. However, this method has
potential application for work packages such as transportation of materials, earthmoving and compaction, and pressure vessel welding.

Off-Site. Off-site estimates of productivity utilize data base information on productivity that has been generated from past company

experience or which has been made available from manufacturers of construction equipment. One approach is to maintain data on production assuming ideal conditions: job site is roomy, trafficability of base is excellent, there are no delays, management and job organization is excellent, operator efficiency is 100%, and work is underway for 60 minutes every hour. This data is then adjusted by shortening the work hour to reflect lost time due to operator breaks or other delays. Ten or fifteen minutes per hour is not uncommon for these breaks. Then, another multiplier is used to reflect the combined degradation effect of other factors such as a poor operator or unfavorable site conditions. This may be as low as 0.5 for the poorer combinations of factors.

Selecting Production Equipment

The choice of production equipment will be influenced by several factors

Volume of Work to Be Performed. Obviously, the size of the job will influence both individual size and numbers of production units.

Job Conditions. As described in Chap. 9, there are many designs of construction equipment, each one best suited for certain job conditions.

Current Inventory of Equipment. The contractor will endeavor to use any available owned or on-hand leased equipment.

Availability of Equipment for Purchase, Lease, or Rental. Some new equipment may be available off-the-shelf from dealers while other items have long leadtimes associated with their procurement; this is especially true of high-capacity equipment. Financing of equipment purchases may be a factor; some equipment cannot be leased or rented locally.

Future Use for Equipment. If equipment will be amortized on a single job, there is not too much concern over its future use. However, if the job will utilize an item only a fraction of its economic life, the contractor must evaluate future use or disposal of the item.

Project Schedule. The contract for the work may include very tight deadlines which force use of larger numbers or sizes of production equipment.

Selection of Support Equipment

Support equipment must be present when needed and have the proper capacity for the work involved. For example, on earthmoving projects there must be enough pushdozers to service the scrapers and enough graders, compactors, water distributers, and similar items to process the material delivered by the production units. In the case of lifting equipment, planners must decide from among rubber-tired, crawler, and tower-mounted items and within the rubber-tired category choose either truck-mounted, self-propelled, or towed units. On large projects, decisions must be made concerning provision of construction gas and compressed air—whether to use a central source and distribution system or to rely on separate units. In selecting the equipment, the estimator must be careful to isolate short-term requirements from the more continuing requirements of the job. For example, a crane may be needed daily on a work package to handle a variety of lifts of less than 20 tons but on two occasions there are lifts of 100 tons. The estimator would probably not select a 100-ton crane for continuous use since it would be underutilized most of the time. He or she would select a 20-ton crane for continuous use and plan to rent a 100-ton crane for the two special lifts. Of course, an economic study of the two alternatives would be necessary to prove this cost effective. The objective, as always, is to keep costs down while providing the support needed to insure that production labor and production equipment are not delayed.

Selection of Administrative Equipment

Each construction firm will tend to have a standard family of company owned administrative equipment to support various types and sizes of jobs. For estimating purposes, the estimator will probably use a checklist that identifies these items and will select types and numbers from this list. Typical administrative equipment items are

Passenger car, light	Ambulance
Passenger car, heavy	Truck, prime mover, 10 ton
Pickup, $\frac{1}{2}$ ton	Truck, stake and platform, 5 ton
Pickup, $\frac{3}{4}$ ton	Trailer, low boy
Pickup, four-wheel drive	Trailer, tilt-bed
Pickup, crew cab	Lubricator, truck-mounted
Scooter	Fuel truck
Carryall	Trailer, office
Van	Forklift, warehouse
Station wagon	Forklift, rough terrain

Some of these items will remain on the project and will be charged off to the project on a weekly or monthly basis. Others are occasional use items and will be charged on an hourly or daily rate. Since administrative equipment is not directly associated with any work package, their costs will be handled as a job indirect cost.

Basic Approach to Developing Equipment Packages

Chapter 13 will discuss the overall approach to developing crews and costs. For purposes of this chapter realize only that a given crew must be a balanced combination of people and equipment if it is to perform its work without excessive idle time on the part of either labor or equipment; later, we will apply this information to the total picture.

The construction schedule is directly related to the number and types of production equipment selected. On the one hand, the equipment list can effectively dictate the schedule; on the other hand, the equipment list can be dictated by the schedule. In the first case, the contractor is not under scheduling pressure, selects crews and equipment conveniently available, applies them to the job and the work progresses at whatever their rate may be. In the second case, a schedule dictates that a work package must be accomplished by a certain time so the resources applied are those necessary to meet that target. In actual practice, both approaches are used simultaneously most of the time in that the estimator will tentatively select crews, determine duration of the work package, and then check to see if this satisfactorily meets scheduling requirements. If not, the estimator will adjust the crews to get other productivities and other durations.

If equipment is the production item (as opposed to labor), the quantity of a pay item is the first item of information needed. Next, the estimator must determine the average productivity of the production equipment under the conditions of the job as previously discussed. Dividing quantity by productivity rate gives the number of equipment days (or hours) to perform the task. The estimator then determines the number of pieces of production equipment to assign to the job. Equipment days (or hours) divided by the number of pieces yields the duration of the work package. As stated before, if this duration is unsatisfactory, the estimator can choose a different number of pieces and adjust the duration. However, the balancing of different production items within an equipment fleet must also be considered.

Once the lists of production equipment have been established, support equipment is added with a capability to match or exceed that of the production items.

Balancing Equipment

On certain work packages, several items may be considered as production equipment. For example, if borrow material must be dug from an embankment with a power shovel, loaded on trucks, delivered to a fill area, spread, and compacted, there are two primary production pieces: the power shovel and the trucks. These must be matched to get the greatest production of a pay item: excavation cubic yards. Support items in the form of graders, compactors, water distributors, etc., will then be matched to these primary production items. Optimizing an equipment fleet is called *equipment balancing.* This technique is shown utilizing an example of a piece of earth loading equipment that is loading trucks for hauling of material to a fill area. The objective is to select the proper number of trucks to match the loading equipment. The determination will be made on the basis of apparent least cost per cubic yard to move the material.

At the outset, you should realize that the organization of the loading area has considerable effect on the outcome. Figure 11-1 illustrates two basic relationships between loader and trucks. In Situation A the

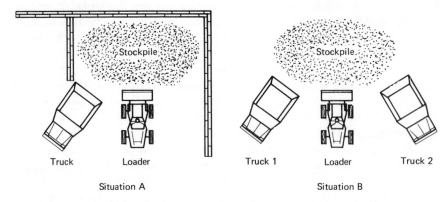

Fig. 11-1   Organization of the loading area.

stockpile is confined by walls so as to restrict truck access to one side of the loader. This means that the loader must wait on each truck as it positions itself for loading. In other words, its least cycle time is the time it takes to load a truck plus time spent in waiting for the truck to get into position. In Situation B, the stockpile is not confined in any way so that trucks can position themselves on either side of the loader. Thus, while Truck 1 is being loaded, Truck 2 can get into position and the loader can move from Truck 1 to Truck 2 without delay. In Situation B, the loader's minimum cycle time is only the time required to load a truck. The time required for the loader to load a truck in both situations will depend on the distance between and angular relationship of the loader and truck as well as the capacities of both pieces of equipment. An arrangement approximately as shown will minimize loader movement. You may wish to visualize other arrangements and their effect on loading time.

Next, we move to the actual equipment balancing actions. Three steps are involved.

1. For each item of equipment, list those steps that can occur during performance of its work. For example

Loader:	Wait for truck
	Dig material
	Maneuver to dump
	Dump
	Return maneuver
	Dig material
	(Repeat above until truck is full)
Dump Trucks:	Wait for turn to be loaded
	Position for loading
	Receive load
	Travel to dump
	Position for dumping
	Dump
	Return travel

2. For the action items (as opposed to the waiting items) determine average times.

3. Balance the items to optimize equipment use.

---

example:

A large stockpile of material is to be transported to a fill area. The contractor has a rubber-tired front loader and a number of dump trucks available for the job. The problem is to determine the optimum number of dump trucks to assign to the job. Based on a study of the job and knowledge of operators, time factors for various steps have been determined and are given below along with cost and capacity factors. The organization of the loading area is similar to that illustrated in Situation A, Fig. 11-1.

Front Loader:	Cost per hour with operator	$40
	Total time to load truck	4 minutes
	(not including positioning time)	
Truck:	Cost per hour with driver	$24
	Positioning time	1 minute
	Loading time	4 minutes
	Travel to dump	6 minutes
	Dump time	2 minutes
	Return travel	5 minutes
	Truck capacity	6 CY

On a piece of paper scaled for time (see Fig. 11-2), we will record in bar graph fashion the activities that would occur under possible combinations. We need study only two options, one which optimizes the loader, the other which optimizes the trucks. We find those options by dividing the truck cycle time by the loader cycle time. In this case, the trucks have a total load, travel, dump, return, and position time of 18 minutes. The least cycle for the front loader is five minutes (4 minutes load time plus 1 minute waiting for truck to get into position). Dividing one by the other gives 3+ meaning that either three or four trucks should be used. As you will see later, three trucks optimizes the trucks, while four trucks optimizes the loader. Figure 11-2 provides graph solutions for three and four trucks.

Using three trucks, the trucks will be in constant motion (theoretically) and only the front loader will have extra idle time. In addition to its normal one minute wait as each truck gets into position, the loader will have 3 additional minutes of waiting after every three truck loads.

Using four trucks, the loader will not have any delays in addition to truck positioning time but the trucks will experience a 2 minute wait at the end of each run. Note that this affects the overall cycle time by adding 2 minutes to it. With the three-truck option, the overall cycle time for the fleet equalled the truck cycle time because the trucks were optimized. Now, the cycle time numerically equals the number of trucks used multiplied by the loader cycle time because the loader is optimized.

To compare the two options, the cost per cubic yard under each is calculated. For three trucks and one loader

$$
\begin{array}{lr}
\text{Cost per hour: Front loader} & \$\ 40 \\
\text{Three trucks} & \underline{\quad 72} \\
& \$112
\end{array}
$$

$$
\text{Volume moved/hour} = \frac{60 \text{ minutes/hour}}{18 \text{ minutes/cycle}} \times (3 \text{ trucks})
$$

$$
\times\ (6 \text{ CY/truck}) = 60 \text{ CY/hour}
$$

$$
\text{Cost/CY} = \frac{\$112/\text{hour}}{60 \text{ CY/hour}} = \$1.87/\text{CY}
$$

Using four trucks with the front loader, the cycle is 20 minutes for each 24 cubic yards, and the cost per hour is now $136 due to the addition of one truck.

$$
\text{Volume moved/hour} = \frac{60 \text{ minutes/hour}}{20 \text{ minutes/cycle}} \times (4 \text{ trucks}) (6 \text{ CY/truck})
$$

$$
= 72 \text{ CY/hour}
$$

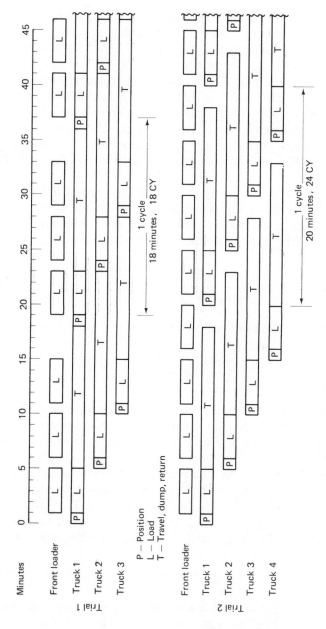

Fig. 11-2 Graphing of equipment cycles.

P — Position
L — Load
T — Travel, dump, return

$$\text{Cost/CY} = \frac{\$136/\text{hour}}{72 \text{ CY/hour}} = \$1.89/\text{CY}$$

The three-truck option is slightly more economical, but not conclusively, in view of the accuracy of assumptions in the problem. Looking back at the productivity calculations, you note that 72 CY per hour can be moved with four trucks while the rate is 60 CY per hour with three. This means that the job can be completed more quickly with four trucks and there may be some savings that result from this that overcome the slightly higher cost per cubic yard as it was calculated. Options other than three or four trucks would not be tested. The loader can load only 72 CY of material per hour and this is done with four trucks; adding other trucks to the job will just create longer delays for all trucks and increase costs.

This example is extremely simple but does illustrate the principle involved. No consideration was given to breaks during the day which effectively reduce each working hour to 45 or 50 minutes. Handle this as follows (assuming front loader plus four trucks and a 50-minute hour):

$$\text{Hourly output} = \frac{50 \text{ minutes}}{20 \text{ minutes per cycle}} \times 24 \text{ CY/cycle} = 60 \text{ CY/hour}$$

$$\text{Cost per hour (as before)} = \$136$$

$$\text{Cost per CY} = \frac{\$136}{60 \text{ CY}} = \$2.27/\text{CY}$$

If we can assume that the time factors used in the problem reflect typical job conditions we will make no further productivity reductions. If there were other inefficiencies yet unaccounted for, productivity would be reduced further.

What we have done in this problem is develop a balanced equipment package. If the contractor had two or more front loaders and additional trucks, two or more of these packages could be assigned to the job and correspondingly higher production rates could be obtained assuming the job could be organized to keep each package from interfering with the other.

Figure 11-3 depicts another situation which requires equipment balancing. It involves the very common problem of assigning enough pushdozers to a fleet of scrapers so that the scrapers are not delayed,

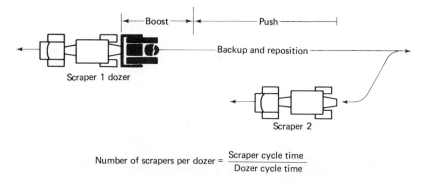

$$\text{Number of scrapers per dozer} = \frac{\text{Scraper cycle time}}{\text{Dozer cycle time}}$$

Fig. 11-3   Balancing scrappers and pushdozers.

yet the dozers are not excessively idle. Scraper cycle time consists of loading, travel to the fill area, spreading, return to cut area, and any waits or delays. Pushdozer cycle time consists of push time while the scraper cuts, boosting time while it helps the scraper accelerate after it is loaded, repositioning time to get behind the next scraper, and any delays. Note the formula in the figure for determining the number of scrapers that can be serviced by one dozer. For example, if the scrapers have a cycle time of 10 minutes and the dozer cycle time is 1.5 minutes, the formula yields a figure of 6.67 scrapers. This indicates that six scrapers can be serviced without scraper delay although the dozer will have some delay. If seven scrapers are assigned, there will be scraper delays. The calculations also allow us to conclude that two pushdozers could service 13 scrapers (2 × 6.67 = 13+).

SPECIAL CONSIDERATIONS INVOLVING EARTHWORK

Soil in its natural state has a certain density. When this soil is disturbed by digging equipment it loosens and has less density. This soil may then be spread and artificially compacted in a new location. The new density may be greater or less than that of its natural state. Soil in its natural state is measured in *bank cubic yards* or *in-place cubic yards*. Loosened material is measured in *loose cubic yards* while artifically compacted material is measured in *compacted cubic yards*. The abbreviations BCY, LCY, and CCY, respectively, apply. Figure 11-4 depicts the relationships among these measures.

It is important in earthwork calculations to know whether you are dealing with BCY, LCY, or CCY in a given operation. An excavator removes BCY, a truck hauls LCY, and a fill contains CCY. Some other terms to learn in connection with this subject are

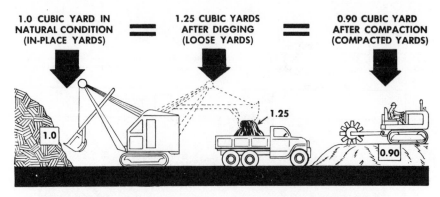

1.0 CUBIC YARD IN
NATURAL CONDITION
(IN-PLACE YARDS)

1.25 CUBIC YARDS
AFTER DIGGING
(LOOSE YARDS)

0.90 CUBIC YARD
AFTER COMPACTION
(COMPACTED YARDS)

1.25

1.0

0.90

Fig. 11-4   Volume relationships for one type of material. [Courtesy US Army (TM 5-252)]

$$\text{Load factor } = LF = \frac{\text{Bank cubic yards*}}{\text{Loose cubic yards*}} = \frac{\text{lb/LCY}}{\text{lb/BCY}}$$

$$\text{Shrinkage factor } = SF = \frac{\text{Compacted cubic yards*}}{\text{Bank cubic yards*}} = \frac{\text{lb/BCY}}{\text{lb/CCY}}$$

Swell = Percentage increase in volume from bank to loose condition

$$= \left(\frac{\text{Loose cubic yards*}}{\text{Bank cubic yards*}} - 1 \right) \times 100\%$$

$$= \left(\frac{\text{lb/BCY}}{\text{lb/LCY}} - 1 \right) \times 100\%$$

*For a given weight of material

Examination of these formulas shows that the load factor relates loose and bank cubic yards and will always be less than 1.0. The shrinkage factor relates compacted and bank cubic yards. If a material is compacted to a greater density in the fill area than it was in bank condition, the SF will be less than 1.0; if less than bank density, it will exceed 1.0. Table 11-1 gives weights and load factors for many common materials.

SPECIAL EQUIPMENT

Transport of Equipment To and From Work Site

The movement of equipment to and from the job site is another cost item to be included in a cost estimate. It involves time, equipment hours, and operator hours. Some equipment is mobile over public high-

TABLE 11-1. Common Material Weights and Load Factors

APPROXIMATE WEIGHT* OF MATERIALS

Material	lb/lcy	(kg/lm³)	lb/bcy	(kg/bm³)	Load Factors
Bauxite	2400	(1424)	3200	(1899)	.75
Caliche	2100	(1246)	3800	(2255)	.55
Cinders	950	(564)	1450	(860)	.66
Carnotite, Uranium Ore	2750	(1632)	3700	2195)	.74
Clay: – Natural bed	2800	(1661)	3400	(2017)	.82
Dry	2500	(1483)	3100	(1839)	.81
Wet	2800	(1661)	3500	(2076)	.80
Clay & Gravel: Dry	2400	(1424)	2800	(1661)	.85
Wet	2600	(1543)	3100	(1839)	.85
Coal: Anthracite, Raw	2000	(1187)	2700	(1602)	.74
Washed	1850	(1098)	2500	(1483)	.74
Bituminous, Raw	1600	(949)	2150	(1276)	.74
Washed	1400	(831)	1900	(1127)	.74
Decomposed Rock:					
75% Rock 25% Earth	3300	(1958)	4700	(2789)	.70
50% Rock, 50% Earth	2900	(1721)	3850	(2284)	.75
25% Rock 75% Earth	2650	(1572)	3300	(1958)	.80
Earth – Dry Packed	2550	(1513)	3200	(1899)	.80
Wet Excavated	2700	(1602)	3400	(2017)	.79
Loam	2100	(1246)	2600	(1543)	.81
Granite – Broken	2800	(1661)	4600	(2729)	.61
Gravel – Pitrun	3250	(1928)	3650	(2166)	.89
Dry	2550	(1513)	2850	(1691)	.89
Dry ¼"-2" (6-51 mm)	2850	(1691)	3200	(1899)	.89
Wet ¼"-2" (6-51 mm)	3400	(2017)	3800	(2255)	.89
Sand & Clay – Loose	2700	(1602)	3400	(2017)	.79
Compacted	4050	(2403)	–	–	–
Gypsum – Broken	3050	(1810)	5350	(3174)	.57
Crushed	2700	(1602)	4700	(2789)	.57
Hematite, Iron Ore, High Grade	4000-5400	(1810-2450)	4700-6400	(2130-2900)	.85
Limestone – Broken	2600	(1543)	4400	(2611)	.59
Magnetite, Iron Ore	4700	(2789)	5500	(3263)	.85
Pyrite, Iron Ore	4350	(2581)	5100	(3026)	.85
Sandstone	2550	(1513)	4250	(2522)	.60
Sand – Dry, Loose	2400	(1424)	2700	(1602)	.89
Damp	2850	(1691)	3200	(1899)	.89
Wet	3100	(1839)	3500	(2077)	.89
Sand & Gravel – Dry	2900	(1721)	3250	(1928)	.89
Wet	3400	(2017)	3750	(2225)	.91
Slag – Broken	2950	(1750)	4950	(2937)	.60
Stone – Crushed	2700	(1602)	4500	(2670)	.60
Taconite	3600-4200	(1630-1900)	5200-6100	(2360-2770)	.58
Top Soil	1600	(949)	2300	(1365)	.70
Traprock – Broken	2950	(1750)	4400	(2611)	.67

*Varies with moisture content, grain size, degree of compaction, etc.
Tests must be made to determine exact material characteristics.
Courtesy Caterpillar Tractor Co.

ways and will travel to the project under its own power. Other equipment must be shipped by truck-trailer, rail, or barge. Transportation expenses are handled as a job indirect cost in that they are part of mobilization and demobilization costs. These costs are extremely easy to overlook or underestimate.

Setting Up Construction Support Plants

On the larger projects, a contractor will choose to set up quarries, rock crushers, concrete batch plants, asphaltic concrete plants, automatic welding jigs, etc. The cost of these efforts must also be accounted for as a job indirect expense.

SUMMARY

As you read this and other chapters it may appear that the estimator has the job of organizing the project since it is stated over and over that the estimator selects the resources for the job. In most cases, the estimator will not be in charge of the job and will not organize crews; this will be the responsibility of the construction manager. However, the estimator's choices of labor and equipment must very closely, if not fully, agree with those of the construction manager or the estimate's validity must be questioned. Thus, when we speak of the estimator organizing work package crews, we really mean the estimator *in coordination with the construction manager*. Some construction firms have a practice of requiring their construction managers to head the estimating team for all projects on which they are the intended construction manager. The reasoning behind this is obvious. In any event, estimators must be thoroughly familiar with construction and the construction firm they represent if they are to realistically estimate a project.

REVIEW

Exercises

1. For the stockpile loading problem described in the chapter, what number of trucks should be used if the loading area organization is similar to Situation B, Fig. 11-1, rather than Situation A?

2. A 6-CY power shovel is being used in conjunction with 18 CY off-highway trucks to transport material from an embankment to a fill area. The following factors apply

    Power Shovel

Time to load one truck	2 minutes

    There is no waiting time for truck positioning if sufficient trucks are available because trucks can be loaded from either side of shovel (trucks back in)

Cost per hour with operator	$100

    Trucks

Time to back into position for loading	1 minute
Loading time	2 minutes

Travel to dump	6 minutes
Dump time	1 minute
Return travel	5 minutes
Cost per hour with driver	$70

Required: Determine the number of trucks to be used for a balanced fleet with least cost per CY.

3. You must excavate 4000 BCY of material and transport it to a fill area. It has a bank weight of 3800 pounds per cubic yard. In its loosened condition it has a weight of 2900 pounds per cubic yard. Its shrinkage factor is 0.95.
   a. What is the load factor?
   b. What is the swell?
   c. What is its compacted density?
   d. If the trucks have a heaped capacity of 8 CY each, how many round trips will be required to move the material?

Questions in Review

1. Distinguish between production equipment and support equipment.
2. Give three examples of administrative construction equipment.
3. List those things that will affect the productivity of a rubber-tired front loader.
4. What is the difference between on-site and off-site productivity determinations?
5. In what way is the job schedule related to the equipment fleet?
6. Why might the least direct cost per pay item equipment fleet not be the least cost fleet?
7. What is the relationship between bank cubic measure and the capacity of trucks, scrapers, and shovels?
8. Are swell and shrinkage factors the same?
9. Give an example of a work package where labor is the production element. Where equipment is the production element.

Discussion Questions

1. The swell factor for a given material is relatively constant. Why isn't the shrinkage factor, or is it?
2. You are going to borrow material from a small hill near the project site by excavating with scrapers. How would you organize the job for maximum efficiency?

SUBCONTRACTING
AND PURCHASING

12

INTRODUCTION

At this point in the text you have been exposed to reasonably detailed coverage of labor and equipment, two of the resources of construction which must be accounted for in any estimate. Materials and installed equipment are other resources. In addition, there may be subcontracts and there is always job overhead and general overhead expenses to consider. Finally, risk and profit must be converted into a markup. In this chapter, the subjects of subcontracting and purchasing will be discussed as background for Chaps. 13 through 15 which will complete discussion of all bid components and wrap up the overall bid preparation process.

Subcontracting and purchasing are the two major functions of procurement. With a subcontract, the general contractor deals with a subcontractor who provides services on the project site in addition to any materials or equipment appropriate to the subcontract. By means of a purchase the general contractor deals with a vendor who delivers a product to the project site but does no work on the site. Thus, the difference between a subcontractor and a vendor hinges upon whether on-site work is performed or not.

SUBCONTRACTS

In general construction, subcontracting of portions of the work is an efficient and economical way of obtaining goods and services in those specialty areas where the general contractor does not maintain a force account capability, in those situations where the general contractor's force account capability is currently committed on other work, or when a subcontractor can perform the work or service more economically.

A subcontract is between a general contractor and another contractor. The client is not a party to this contract. Responsibility for performance of the subcontractor rests with the general contractor. However, the client may, in the original contract with the general contractor, reserve the right to approve all subcontractors by name, or in some cases (such as cost-plus contracts) review and approve each subcontract before subcontract signature.

A contractor intending to bid on a construction project will, during his or her review of the plans and specifications for the project, isolate those construction work packages or services which are suitable for subcontracting. For construction works, the contractor will extract those drawings and specifications which apply to each package. (You will recall from Chap. 4 that a section of a specification is the smallest biddable element in a contract document package; a well-prepared contract document package will start each section on a new page to permit their easy withdrawal for subcontracting.) The contractor prepares subcontract document packages very similar to those described in Chap. 4 for primary contracts and invites bids from apparently qualified bidders. The general contractor will not award any subcontracts unless he or she gets the general contract. The bidding documents will request bids which are valid until after the opening of bids on the general contract.

The subcontract work becomes a work package within the project and its schedule must be integrated into the overall schedule. In all likelihood, the contract documents establish milestone dates for start and completion of the work to insure this integration.

Subcontract work is treated as a direct cost item if its product becomes part of the permanent facility. It is treated as an indirect cost item if its purpose is to develop construction support facilities (such as the project office building) or to provide administrative support services (such as portable toilets). In either case, the general contractor will include an amount in overhead to reflect the costs of subcontract preparation and administration. This will be covered in more detail in Chap. 14.

One of the great advantages of subcontracting is that it reduces the risk borne by the general contractor. The general contractor will have firm dollar bids on the subcontracted portions of the work before submitting a bid as a general contractor and so must prepare detailed estimates only on the remaining force account work. As the extreme case, recall from Chap. 2 that one form of contracting becoming more common has the client engaging in a hard dollar contract with a construction manager (CM) who has no force account capability at all. Thus, 100% of the work is subcontract and the CM gives maximum spread to the risk. Of course, the CM is not without risk. The CM can never be sure that all subcontractors will perform satisfactorily, yet is entirely responsible for their work as far as the client is concerned.

There are at least three risks with subcontracting. The obvious one is that the subcontractor will default and the general contractor (or bonding company) must take over the work. A second is that the subcontractor cannot perform at the required rate and this delays the entire project, again requiring general contractor assistance or takeover. A third risk is that there is a misunderstanding between the general contractor and subcontractor on the total inclusive elements of the subcontract. This risk can be reduced only by thorough preparation of subcontract documents and by use of prebid conferences with potential subcontractors to explain the work and resolve all questions.

PURCHASES

Purchases may be for materials, equipment, or services. Formal purchasing involves the use of purchase orders or contracts. However, the process may be informal as when a small contractor deals over the counter on a cash or charge basis for off-the-shelf items with documentation no more formal than a sales slip. Some purchases require documentation comparable to a construction contract such as those for engineering, fabrication, and delivery of a special piece of equipment to be installed in a project.

Materials

Materials required in large amounts may be purchased through contracts obtained by the competitive bidding process in the manner described for subcontracts. For small amounts, quotations may be obtained informally by telephone or by presenting a list of requirements to several potential suppliers and asking for their quotation without other bidding formalities. In some cases a construction firm will have open-end contracts with suppliers for the furnishing of materials and these materials are available to any job at the preestablished prices during the period of the contract. Many items will be ordered from standard catalogs so catalog prices are used in the estimate. There is always the possibility that quotations cannot be received on material items prior to submission of a bid so the estimator must base all estimates on past experience, commercial cost guides, or whatever information is available at the time. Companies with formal procurement departments will probably maintain costing guides for use by estimators which list current costs of common items, ranges of costs for specification equipment, anticipated lead times for procurement, and expected cost changes in the future.

Material requirements for permanent facilities are identified by measurement or counting during the *quantity take-off* (quantity survey) phase of the estimating process. In addition, the contractor must determine what construction support structures will be needed on the project and develop materials lists for those (project office, warehouses, haul roads and bridges, etc.). The estimator also must account for those supplies which are not part of a structure but are expended in supporting construction or in project administration.

The cost of materials used in the estimate normally is the total cost for delivery to the project site, including base material price, transportation, and taxes. Some contractors choose to separate the tax portion and handle it as an indirect cost; this is done because tax rates vary from state to state and, on public projects, there is a tax exemption. Thus, inclusion of taxes in materials costs could be misleading in future use of the data. All costs of handling the materials on the project site after delivery will be incorporated in the warehousing and storage operations (job indirect cost) or in the labor and equipment cost associated with their use (direct cost). Profit is not included in the materials cost; it is handled strictly in the markup stage (Chap. 14).

For estimating purposes, the estimator will find it convenient to subcategorize materials as follows:

Permanent Materials. Materials, such as concrete, reinforcing steel, piping, etc., which are incorporated into the structures being built for the client are permanent materials.

Materials and Supplies. Material items which are used or expended in the installation of permanent materials, but not incorporated therein, are in this category. Form lumber, explosives, drill bits, surveying stakes, and saw blades are typical materials and supplies. In addition, any materials involved in the construction of job indirect structures can be included in this category.

Installed Equipment

Installed equipment refers to equipment such as escalators, air conditioning systems, generators, and comparable equipment which is installed in a permanent structure. On some projects the procurement of these items is left to the general contractors and they order them on the basis of specifications for the items included within the contract documents. On other projects, the client retains responsibility for the ordering of certain major pieces of installed equipment. For example, on a power plant, the client normally orders the steam supply system, turbine-generators, and other long leadtime major items, leaving the procurement of other items of installed equipment to the general contractor.

If the item of equipment is fully designed and the contract documents contain all necessary drawings and specifications, the procurement of the item is essentially the same as for subcontracting for construction work or services—the general contractor prepares a bid package using extracted design drawings and specifications and invites bids. On turnkey contracts the engineer may choose to subcontract both the design and the furnishing of the item. In such cases the requirements must be described in terms of performance expected and features to be incorporated. It is unlikely that a firm price bid would be received unless the vendor has stock designs that meet the specifications. However, this is not normally a problem since most turnkey work is cost-plus.

In preparing cost estimates, the cost for installed equipment should be the total cost delivered to the project site. Their cost will be a direct cost item.

Services

The provision of portable toilets, first aid facilities, guard service, survey support, and materials testing are examples of project support services. Some of these will be handled through subcontracts or purchase orders, others will be provided by the general contractor. All are job indirect cost items.

The Inflation Problem in Procurement

Most clients prefer to handle construction on a fixed-price basis rather than cost-plus for the obvious reason that costs are defined at the beginning of the contract rather than at the end. Similarly, a contractor bidding on a fixed-price contract seeks to obtain firm prices on all materials and installed equipment needed for the contract. Unfortunately, and especially in times of significant inflation, vendors and suppliers are reluctant to provide fixed quotations. Instead, they want any supply contract to contain a clause which states prices will be those in effect at time of delivery. Obviously, this puts a contractor in a difficult position. One solution is to simply inflate all bid costs using a best judgment of inflation rates that will prevail. A second solution is to seek inclusion of escalation clauses in the contract to cover critical items. Such clauses stipulate that a bid is based upon specific costs for critical items but will be automatically adjusted for changes in these prices beyond a certain amount. A third solution is for the contractor to stockpile needed materials immediately upon award of the contract. Of course, this option must consider the additional costs of interest on money tied up, storage losses, and warehousing costs and include those in the bid. If this option is chosen, the contractor should seek client agreement to pay for materials upon purchase rather than upon installation. A fourth option is for the contract to be rewritten to have the client purchase the materials. Obviously, this shifts considerable risk to the client, but this may be an acceptable trade-off for a bid which does not include high contingency markup for inflation. Should any of these protective approaches not be possible the contractor should at least have a written understanding with the client that bid prices assume project completion by a certain date and that any delays attributable to the client will be basis for adjustment of material prices within the bid.

SUMMARY

In the preparation of an estimate, the estimator must first identify requirements for permanent materials and installed equipment. Later, he or she will determine requirements for materials and supplies and any support services. All of these must be costed. When possible, the estimator will use subcontractor and vendor quotations, open-end contract prices, or catalog prices. If none of these are available in time to incorporate in the bid, the estimator must use any sources available, including experienced judgment, to arrive at a cost figure. Not to be forgotten is the possibility of inflation of materials and equipment costs. If subcontracting will be used, every effort must be made to obtain firm bids prior to bid submission.

REVIEW

Questions in Review

1. What does procurement encompass?
2. What is the difference between a subcontractor and a vendor?
3. Compare the documentation used for primary contracting and for subcontracting.
4. Give the advantages and disadvantages of subcontracting.
5. What is the difference between permanent materials and materials and supplies?
6. Can the same type item be both a direct and an indirect cost item?

Discussion Questions

1. Why should so much fuss be made over distinguishing material items as direct and indirect?
2. Does subcontracting add to or detract from overall project quality?
3. Why would a client take the responsibility for procurement of items of installed equipment rather than pass this on to the general contractor?

DETERMINING WORK PACKAGE RESOURCE REQUIREMENTS

13

INTRODUCTION

Previous chapters have discussed in detail the resources of construction. A major task of the estimator is to determine as closely as possible the quantities of these resources that will be required for each work package and to translate these requirements into work package costs and durations.

The drawings and specifications for a project will yield two of the five resources of construction directly: permanent materials and installed equipment. This is done through a *quantity take-off* where quantities are determined by actual count and measurement from the drawings. Three resources of construction (other materials and supplies (M&S), labor, and construction equipment) are not found on the drawings. However, these can be the high cost items for many work packages and it is essential that they be carefully estimated. For relatively simple construction or repetitive type of work, an estimator may be able to reasonably account for employee-hours and costs associated with these latter three resources by applying factors from handbooks to the quantities of permanent materials and installed equipment. However, this approach can be dangerous in that it is easy to overlook cost items or project features which make handbook data erroneous. The more complete approach, and the one recommended by this text, is the preparation of detailed work package collection sheets which catalog, code, quantify, and cost all five resource categories. While this approach is definitely more time consuming initially, it is the only approach which will establish a basis for close control of resources during construction. Ultimately, it will result in easier preparation of estimates since it leads to development of experience data which will prove far more accurate than any commercial handbook. If the company's

212

estimating system is computerized, it is possible to develop the data base so that the computer can produce printed checklists of work package resource requirements from past projects (or present them on a CRT screen) for use in future estimate preparation.

The counting of permanent materials and installed equipment from the drawings and specifications requires complete understanding of blueprint reading since symbols and abbreviated notation are utilized on all drawings. Proficiency in blueprint reading can be gained only through considerable practice so no attempt will be made to develop that subject in this text. In fact, the specialized nature of each construction discipline (civil, mechanical, electrical, etc.) causes most large estimating offices to group their estimators by discipline since obtaining proficiency in all is not practical nor usual.

The determination of labor, M&S, and construction equipment requirements requires extensive knowledge of construction methods and equipment. Again, expertise here requires considerable experience. In this text, only the outline procedures for development of work package collection sheets can be presented. The presentation will be handled in three ways. First, the subjects of earthwork, concrete, and carpentry will be discussed in some detail to include measurement procedures and the preparation of example work package collection sheets. Next, various other craft operations—masonry, structural steel, pressure vessel piping, other piping, electrical installation, mechanical installation, base courses, and paving—will be discussed in lesser detail but with enough coverage to outline work methods and items to consider in estimating representative work packages. These work items have been selected since they are among the more common ones which a large general contractor will accomplish with force-account resources. Finally, brief mention is made of those work tasks which are almost always subcontracted to specialty contractors on a fixed-price bid basis.

EARTHWORK

It is extremely difficult and costly to obtain detailed maps of the ground surface configuration or the boundaries between various layers of materials in the earth's crust for use in estimating. Earthwork is normally estimated using surface maps that have been drawn following a line or grid type of survey. The survey establishes elevations at various points in the project area from which a contour map is drawn. Mapping of subsurface formations is accomplished using a similar line or grid approach with borings. In either case, the details between known points must be developed by interpolation so any calculations made from this information is approximate at best.

Two other problems with earthwork measurement are the swell

and shrinkage factors discussed in Chap. 11. Recall that in-place material swells when loosened by digging equipment (see Table 11-1). If this material is then compacted in a new location, the volume is again reduced (shrinkage). The new volume may be greater or less than the original bank volume depending upon the characteristics of the material and the compacting effort applied.

Measuring Earthwork

Earthwork is measured by volume, usually cubic yards or cubic meters. The measurement must be further defined as bank (in-place), loose, or compacted to insure proper accounting of the swell and shrinkage factors.

For earthwork such as basement excavations, where the excavation's shape is some combination of common geometric volumes, the volume is readily calculated using volume formulas for those shapes. For most other earthwork, the *prismoidal formula* provides the best approach. The prismoidal formula is applicable to a volume which has two parallel faces. In Fig. 13-1, faces *abcde* and *a'b'c'e'* of the volume shown are parallel but not of the same outline. Also shown on the

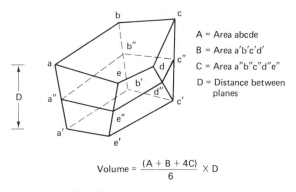

$$\text{Volume} = \frac{(A + B + 4C)}{6} \times D$$

Fig. 13-1   Prismoidal formula.

figure is a plane $a''b''c''d''e''$ which is parallel to and midway between the bounding parallel planes. The prismoidal formula is given in the figure. If one can assume that the area of the midplane is approximately equal to the average of the two end areas, the prismoidal formula can be replaced by

$$\text{Volume} = \frac{(A + B)}{2} \times D*$$

*This assumption creates inaccuracies if A and B are significantly different in area.

Considering other inaccuracies already built into earthwork estimation, the latter formula generally will provide satisfactory results. Two applications are described in this chapter: the *end area method* and the *grid method.*

End Area Method

The end area method assumes that the land to be cut or filled is sliced vertically, the slices being parallel and a fixed distance apart (see Fig. 13-2). The volume of earth between two slices is equal to the average area of the slice between existing and proposed ground lines multiplied by the distance between the slices. Formulas for this are given in the Fig. 13-2.

To illustrate this method by example, assume that an estimator is to determine the earthwork cut and fill in a section of the project shown in Fig. 13-3. The lot size is 300 ft × 300 ft; existing contour lines are solid; proposed contour lines are dashed. To utilize the end area approach the project must be sliced into parts. For this purpose and later uses, we will superimpose a 100-ft square grid network over the project. This is shown in Fig. 13-4. Note also the addition of the dotted line that shows the boundary between cut and fill sections of the project. This line is obtained by connecting the points of intersection of existing and proposed contour lines of the same elevation and will be helpful in keeping calculations straight. Lines A–A' through D–D' define vertical cross-sections to be used in end area calculations.

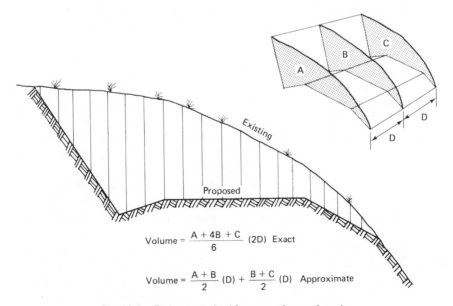

$$\text{Volume} = \frac{A + 4B + C}{6} \ (2D) \ \text{Exact}$$

$$\text{Volume} = \frac{A + B}{2} \ (D) + \frac{B + C}{2} \ (D) \ \text{Approximate}$$

Fig. 13-2   End area method for measuring earthwork.

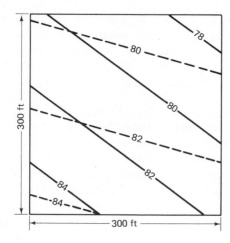

Fig. 13-3   Example earthwork project.

For brevity, we will calculate volumes only between planes B-B' and C-C'. For both of these cross-sections, profiles of the existing and proposed surface shapes are drawn to scale. These are also shown in Fig. 13-4. Next, the areas of the cut and fill portions of each profile are calculated and the modified prismoidal formula is applied to calculate the volume as illustrated in the figure.

Grid Method

With the grid method the area of cut or fill is divided into grid squares over its surface. Each grid line represents a vertical cut into the earth, much like cutting a pan of fudge into squares. The top and bottom surfaces of each piece represent existing and proposed ground surfaces and the distance between them is the depth of cut or fill. This is illustrated in Fig. 13-5.

With the grid method one assumes that the top and bottom of a volume are parallel and equal in area and that depth is measured at the center of each grid between existing and proposed surfaces. Using the same example as before we will calculate the volumes of cut and fill for the same section of the project used in the end area example. Referring to Fig. 13-6, note that the grids being used have been lettered $A1$, $A2$, B, and C. Grid A was divided into two parts because it contains both cut and fill—$A1$ is the cut portion and $A2$ the fill. The next step is to determine the areas and depths to be used. In those grid squares where only cut or only fill is involved, the area is that of the grid: 10,000 SF. It is estimated that grid $A1$ is 7500 SF and grid $A2$ is 2500 SF. As for depth, the procedure is to locate the geometric center of each cut or fill shape, determine the difference between existing and

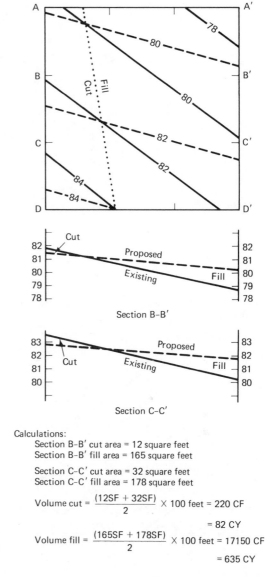

Fig. 13-4 Calculations using end area method.

Calculations:
  Section B-B' cut area = 12 square feet
  Section B-B' fill area = 165 square feet

  Section C-C' cut area = 32 square feet
  Section C-C' fill area = 178 square feet

  Volume cut = $\dfrac{(12SF + 32SF)}{2}$ × 100 feet = 220 CF

                                                    = 82 CY
  Volume fill = $\dfrac{(165SF + 178SF)}{2}$ × 100 feet = 17150 CF

                                                    = 635 CY

proposed elevations at that point and use that as the distance. Since it is seldom that the center of a grid will coincide with the contour lines, it will be necessary to interpolate elevations. Do this by drawing a line that is approximately perpendicular to the two bounding contours and passing through the center point, and then interpolate along this line. See Fig. 13-7 for this procedure. Referring again to Fig. 13-6 and the

Volume = (area of grid) X (distance between existing and proposed surfaces)

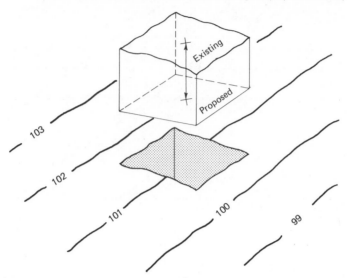

Fig. 13-5   Earthwork calculations using grid method.

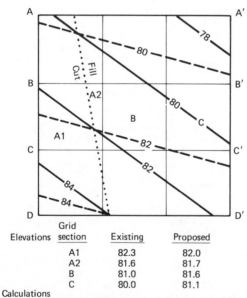

Elevations	Grid section	Existing	Proposed
	A1	82.3	82.0
	A2	81.6	81.7
	B	81.0	81.6
	C	80.0	81.1

Calculations

Cut: Section A1 = (7500 SF)(82.3 − 82.0)   =   <u>2250 CF</u>
                                  Total cut   =   2250 CF

Fill: Section A2 = (2500 SF)(81.6 − 81.7)   =   250 CF
        Section B = (10,000 SF)(81.0 − 81.6)   =   6000 CF
        Section C = (10,000 SF)(80.0 − 81.1)   =   <u>11000 CF</u>
                                  Total fill   =   17250 CF

Fig. 13-6   Calculations using grid method.

218

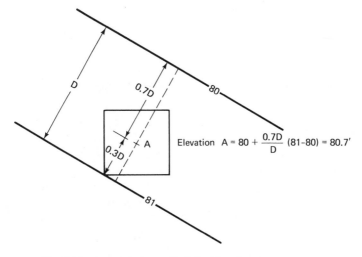

Elevation $A = 80 + \dfrac{0.7D}{D}(81{-}80) = 80.7'$

Fig. 13-7   Interpolating to find elevations between contours.

calculations, it will be seen that the volumes determined by this method are comparable to those derived from the end area. As a general rule, use the end area method for complicated cross-sections, such as road cuts, and the grid method when existing and proposed profiles are relatively straight.

Computerized Earthwork Calculations

A contractor specializing in earthwork operations will generally utilize a computer for the calculation of earthwork quantities since the sophisticated programs available permit rapid and accurate calculations.

Swell and Shrinkage

The quantities calculated in the example are bank for the cut yardage and compacted for the fill sections. To illustrate the handling of swell and shrinkage, we will assume that the material is decomposed rock (25% rock, 75% earth) and that the fill area is being brought to grade with material from the cut area.

From Table 11-1 for decomposed rock

$$\text{Pounds per loose cubic yard } = \text{ 2650 lb}$$

$$\text{Pounds per bank cubic yard } = \text{ 3300 lb}$$

$$\text{Load factor (LF) } = \text{ 0.80}$$

$$\text{Volume of one bank cubic yard after loosening } = \frac{3300}{2650} = 1.25 \text{ CY}$$

If scrapers are used for the cutting operation and they have a heaped capacity of 20 CY, the volume of bank material they can carry in one load is

$$\text{Bank volume/load} = \frac{20 \text{ LCY}}{1.25} = 16 \text{ BCY}$$

or

$$= (20 \text{ LCY})(0.8 \text{ LF}) = 16 \text{ BCY}$$

This figure would then be used in calculating required loads and other scraper production data.

The material cut and hauled by scrapers is spread in the fill area for compaction. As stated before, the amount of compaction possible varies with the material and with the amount and type of compactive effort applied. In this case, assume that the shrinkage factor (SF) is 0.90. That is

$$\text{SF} = \frac{\text{Compacted cubic yards}}{\text{Bank cubic yards}} = 0.90$$

Since each scraper will haul 16 BCY (20 loose yards), it will contribute this amount to the fill

$$\text{Fill} = 0.90 \times 16 \text{ BCY} = 14.4 \text{ CCY/load}$$

Another calculation will be the amount of bank cubic yards necessary to satisfy the fill requirements. Assume 1400 CCY are required.

$$\text{Bank requirements} = \frac{1400 \text{ CCY}}{0.9 \text{ LF}} = 1555 \text{ BCY}$$

Determining Resources and Cost for an Earthmoving Work Package

It has been mentioned a number of times that the planners will break the project into work packages. One or more of these packages could involve earthwork so, at this point, the procedure for itemizing and costing all resource requirements in such a package will be described. Figure 13-8 is a Work Package Collection Sheet on which calculations and entries are made. This figure should be referenced during this discussion. First, note that the Work Package Collection Sheet is a preprinted form, typical of many worksheets that an estimator might use. Its purpose is to organize the itemization of the resources in a work package, their cost, and the work package duration. It also can serve as a coding form for entry into a computer-based estimating pro-

WORK PACKAGE COLLECTION SHEET

System/Structure Identifier: 0 1 . 3 0 0
Crew Level Work Package Identifier: 0 2 . 1 1 1
Description: EXCAVATION, AREA 3, AND DISPOSAL W/O COMPACTION

Productivity

Base Unit for Productivity: BANK CY
Total Quantity Base Unit: 24,000 BCY

Duration (Crew-Hours): | Low | Target | High |
| --- | --- | --- |
| | 32 | |

Escalation Rates (%):
Materials □
Labor □
Equipment □

Notes

L.F. = 0.8
CYCLE = 6.3 MIN; ASSUME 45 MIN HOUR

$$BCY/HR = \left(\frac{45\ MIN}{6.3\ MIN}\right)(22CY)(0.8LF)(6\ SCRAPERS)$$

$$= 754\ BCY/HR$$

$$DURATION = \frac{24,000\ BCY}{754\ BCY} = 32\ HOURS$$

ADD 1 DAY FOR BAD WEATHER
∴ PROJECT WILL LAST ONE WORK WEEK

Cost Summary

PM =
M&S =
IE =
L = 5,398.40
CE = 22,847.20
TOTAL = 28,245.60

Permanent Materials (PM)

Resource Code	Description	Unit	QTY	Unit Cost Low	Target	High	Extension

Other Materials and Supplies (M&S)

Installed Equipment (IE)

Crew Labor (L)

Resource Code	Crew Labor (L)	NR.	Cost/Hour Low	Target	High	Extension
0 8 1 1	FOREMAN	1		13.90		444.80
0 8 1 2	EQUIP OPER, MEDIUM	10		13.40		4288.00
0 8 1 3	SPOTTER	2		10.40		665.00
						5398.40

Equipment Not Charged As Job Indirects

Resource Code	Description	NR.	AVAILABILITY Periods	$/Period	USE Hours	$/Hour	Extension
7 3 2 2	SCRAPER, ELEV, 22 CY	6 @	1 × 1850 -	+ (32) ×	24.00	= 15,708.00
7 2 0 7	DOZER, D7	3 @	1 × 1600 -	+ (32) ×	11.25	= 5,808.00
7 4 1 2	GRADER, 12'	1 @	1 × 1000 -	+ (32) ×	8.0	= 1,259.20
		@	×	+ ×		= 22,847.20

Fig. 13-8 Work Package Collection Sheet for excavation.

gram. At the top, the work package is described in words and with a work package coding identifier. Also, the location of the work is identified through the system/structure identifier. Blank sections are available for listing any of the five resources of construction. For this example, assume that the work package is the cutting of soil with elevating scrapers and the moving of this material to a disposal area where it will be spread by dozers without compaction. Planners have determined that a fleet of six scrapers supported by three dozers and a grader are needed for the operation. This is strictly a labor and equipment work package so no entries are appropriate for other resources. The quantity survey effort has determined that 24,000 BCY are to be cut and moved. Note that this is entered in the upper right section of the form. Just above this entry is another called Base Unit for Productivity. This is the production unit against which crew productivity is measured; the target rate has been determined to be 754 BCY/hour in the "Notes" section using methods described in Chap. 11. In this same section, the duration of the package was found to be 32 hours or 4 days of 8 hours each. In the Crew Labor section the crew is listed, coded, and wage rates assigned. By multiplying duration by numbers of workers and wage rates, the cost figures in the extension column are derived. In this example, it is assumed that the wage rates include base wages plus mandatory insurance/tax items (Chap. 7). Other fringes would be handled as a bulk job indirect. Equipment is handled similar to labor; the cost figures are derived from 32 hours of use charges and an appropriate number of days of availability charges. In this example, an allowance for 1 day of bad weather and 2 days of weekend results in assessment of a full week of availability charge.

Not used on this form were the "low" and "high" spaces in the cost and productivity sections. Also, escalation rates were not included. Those sections could be used in a computerized estimating system of the type to be described in Chap. 16.

CONCRETE

Either plain or reinforced concrete may be placed. Concrete is produced using one of the several types of portland cement or substitutes. Various admixtures may be introduced to impart special characteristics to the mix. Aggregates used with the cement are selected for strength, weight, appearance, or other quality. Structural concrete is normally ordered in terms of compressive strength (psi), cement type, and admixtures.

A concrete operation includes base preparation, waterproofing (if applicable), forming, placement of reinforcing steel and embedments (if applicable), concrete placement, curing, form removal, and finishing.

It is essential that estimators separately consider each type of concrete operation (ground slabs, overhead slabs, foundation walls, columns, beams, etc.) since the complexity of forming and resteel, the employee-hour requirements per cubic yard, and the type of placement equipment used will vary considerably.

Forming

Forms are not shown on the contract drawings. The estimator must anticipate their need and determine their quantity, installation effort and cost using judgment and past experience. Forms may be custom made for the project or commercial fiberglass, wood, paper, or steel forms may be utilized. If locally produced, the form surface is normally of special plywood (plyform) and backed by a framework of dimension lumber or structural steel shapes. The construction of forms for a project is costly because of the materials and labor involved. Whenever possible, a contractor will design forms which can be reused several times to save on these costs. Commercial forms may be purchased or rented and are suitable for repeated use making them particularly attractive to the contractor who tends to do the same type of concrete work over and over.

Form design is one of the engineering tasks facing a concrete contractor. He or she must be able to design forms and shoring which will not yield under the heavy pressure of fresh concrete. Although an estimator will not design forms, he or she must be able to visualize form requirements and translate them into approximate resource quantities and costs.

There are a number of accessories that may be part of the forming operation. Form oil is placed on the surface of the form to prevent bonding to the concrete. Concrete ties and spreaders are bolt-like devices or wires placed in wall forms to keep forms at the specified separation. Various embedments may be attached to the forms which become part of the completed concrete mass (hangars, threaded inserts, column bolts, etc.). An assortment of special clamps may be used with special forms, such as for columns.

An estimator first determines the square feet of contact area required. This quantity is then factored to obtain quantities of materials required. For example, factors for foundation wall forms might be

For each square foot of contact area (SFCA)

Order 1.1 SF of plyform
Order 1.4 board feet of dimension lumber
Add 5% of lumber cost to account for accessories

Factors for other forms can be expected to differ considerably from these, particularly those for dimension lumber. On overhead forms, the estimator must remember to include proper shoring.

Preparing a Collection Sheet for Forming

Figure 13-9 is a completed Work Package Collection Sheet for form fabrication. Separate collection sheets should be used for setting and stripping since multiple use of forms gives different square footage requirements among the various forming subtasks. In this case, the resources involved are M&S, labor, and equipment. There are no permanent materials unless some embedments are required. The total quantity of work on the sample collection sheet is 7800 SFCA and this has been converted by the factors just described to give the quantities to order in the M&S section. An hourly tool charge is included on the form. Productivity rates and costs come from company data files.

Reinforcing

Reinforcing is normally in the form of deformed steel bars (rods with deformation to facilitate bonding), welded wire fabric, prestressed cables or posttensioned cables. Some applications call for steel or glass fibers.

Deformed steel bars come in standard sizes and several grades (Table 13-1), and must be ordered to conform with the drawings and specifications. Typical designs for reinforced concrete call for both straight and bent reinforcing bars. The contractor has the choice of bending bars on the site or custom ordering them from a fabricator. If the contractor chooses to bend them on site, he or she will order bars in

TABLE 13-1.   Reinforcing Bar Sizes

Bar Size	Weight (lb/ft)	Diameter	Area
2	0.167	0.250	0.05
3	0.376	0.375	0.11
4	0.668	0.500	0.20
5	1.043	0.625	0.31
6	1.502	0.750	0.44
7	2.044	0.875	0.60
8	2.670	1.000	0.79
9	3.400	1.128	1.00
10	4.303	1.270	1.27
11	5.313	1.410	1.56
14	7.65	1.693	2.25
18	13.60	2.257	4.00

WORK PACKAGE COLLECTION SHEET

System/Structure Identifier: 0 2 • 1 / 2
Crew Level Work Package Identifier: 0 3 • 3 / 1
Description: FABRICATION OF WALL FORMS

Productivity: SQUARE FEET CONTACT AREA

Base Unit for Productivity: 7800 SFCA

Total Quantity Base Unit: 7800 SFCA

Duration (Crew-Hours)

	Low	Target	High
		72	

Escalation Rates (%)

Materials ☐
Labor ☐
Equipment ☐

Permanent Materials (PM)

Resource Code	Description	Unit	QTY	Unit Cost Low	Target	High	Extension
1 1 0 5	PLYFORM 5/8"	SF	8580		— 52		4461 60
1 2 0 0	DIMENSION LUMBER	BF	10,920		— 30		3276 00
1 8 0 0	ACCESSORIES	ALLOW	—				387 00

Other Materials and Supplies (M&S)

Installed Equipment (IE)

| | | | | | | | 8124 60 |

Crew Labor (L)

	Description		NR.	Cost/Hour Low	Target	High	Extension
0 2 1 2	CARPENTERS		4		13 20		3801 60
0 2 1 3	LABORER		1		10 40		748 80

Equipment Not Charged As Job Indirects

		NR.	AVAILABILITY Periods	$/Period		Hours	USE $/Hour		Extension
7 0 0 2	SMALL TOOLS	@	③	×	+ (72 ×	2.50) =	180 00
		@	③	×	+ (×	) =	
		@	③	×	+ (×	) =	
		@	③	×	+ (×	) =	

Notes

DURATION = 7800 SFCA / 110 SFCA/HR = 71 HOURS

SAY 72 HOURS OR 9 DAYS

Cost Summary:
PM = M&S = 8124.00
IE =
L = 4550.40
CE = 180.00
TOTAL = 12,855.00

Fig. 13-9   Work Package Collection Sheet for forms.

225

mill lengths by size and total weight, prepare shop drawings for the various shapes, and erect jigs for bending. If the contractor deals with a fabricator, the fabricator will prepare the shop drawings, cut, and bend the steel. In either case, the total cost of resteel is a combination of the material price, shop drawings, fabrication, delivery, and placement costs. In addition, there are various accessories such as high-chairs, spacers, tie-wires, clips, etc., that are required in resteel placement. These are accounted for by a percentage markup of resteel costs.

Reinforcing bars are normally available in 20-, 40-, and 60-ft lengths. When continuous reinforcement is specified over greater lengths or different bars meet, the bars must be overlapped by a specified number of diameters (part of codes or contract specifications) to provide this continuity. The extra resteel used in overlapping must be accounted for in the estimate. Alternatively, the bars may be joined by welding (which introduces another labor and equipment cost), or mechanically connected using *cadwelds* which are special-purpose joint sleeves slipped over the butted ends of resteel to be joined. The bond is achieved by pouring a molten metal into the sleeve. The molten metals hardens and forms a joint stronger than the resteel itself. Another form of mechanical joint for resteel requires threading of the ends of the bars and then connecting them with threaded couplings. Still another method employs a steel sleeve slipped over the ends to be joined and compressed hydraulically over the resteel deformations. If cadwelding or other mechanical joining technique is required, estimators should count the number of each size connection required since a different kit is required for each size. Employee-hour requirements for cadwelding vary with the location of the weld since each weld requires special equipment at the site of the weld and this equipment is difficult to handle in awkward positions. Whenever practical, resteel mats and cages are prefabricated on the ground and then hoisted into position—this is both safer and less costly.

Welded wire fabric (or wire mesh) is fence-like material with wires in a grid pattern. Welded wire fabric comes in many sizes including those shown in Table 13-2. Lighter welded wire fabric is shipped in rolls, usually 60 in × 150 ft. Heavier gauges are in sheets. Welded wire fabric is ordered by area, usually square feet (SF) or hundreds of square feet (CSF). In ordering, allowance must be made for approximately one square width for overlap of adjacent rolls or sheets.

A posttensioning cable (tendon) consists of the cable, an enclosing metal or plastic tube, and conical anchors at each end. It is positioned inside the forms with the ends of the cable protruding through the forms. After concrete has been placed and cured, jacks are attached to the free ends and tension applied. When proper tension has been achieved, the ends are anchored (in some cases the tube is filled with

TABLE 13-2. Welded Wire Fabric Sizes

Designation	Wire Spacing		Wire Gauge		Sectional Area		Weight per CSF
	Longit.	Transv.	Longit.	Transv.	Longit.	Transv.	
2 × 2—10/10	2	2	10	10	0.086	0.086	60
2 × 2—12/12	2	2	12	12	0.052	0.052	37
3 × 3—8/8	3	3	8	8	0.082	0.082	58
3 × 3—10/10	3	3	10	10	0.057	0.057	41
4 × 4—4/4	4	4	4	4	0.120	0.120	85
4 × 4—6/6	4	4	6	6	0.087	0.087	62
4 × 4—8/8	4	4	8	8	0.062	0.062	44
4 × 4—10/10	4	4	10	10	0.043	0.043	31
6 × 6—0/0	6	6	0	0	0.148	0.148	107
6 × 6—1/1	6	6	1	1	0.126	0.126	91
6 × 6—2/2	6	6	2	2	0.108	0.108	78
6 × 6—3/3	6	6	3	3	0.093	0.093	68
6 × 6—4/4	6	6	4	4	0.080	0.080	58
6 × 6—4/6	6	6	4	6	0.808	0.058	50
6 × 6—5/5	6	6	5	5	0.067	0.067	49
6 × 6—6/6	6	6	6	6	0.058	0.058	42
6 × 6—8/8	6	6	8	8	0.041	0.041	30
6 × 6—10/10	6	6	10	10	0.029	0.029	21

grout). Jacks are removed, cables cut flush, and finish patching put over the anchor ends.

Concrete

Concrete may be produced on site or supplied from a commercial source. For very small jobs, portable mixers which produce several cubic feet per batch may be used. The cost of this concrete is the delivered cost of the materials (cement, sand, gravel, water, and additives), ownership and operating costs for the mixer, and labor. On extremely large projects, especially those in remote locations or which require highly controlled concrete mixes, the contractor will establish an on-site concrete batch plant. Computerized units which produce in excess of 200 CY/hour are available. For concrete produced from such plants the estimator must determine an equivalent cost/CY. This cost must incorporate erection, ownership and operating, raw material, and disassembly costs for the plant plus all on-site delivery costs (batch trucks and operators).

Perhaps the most common source of concrete today is the commercial batch plant. These plants are capable of producing and delivering all common mix designs on short notice and may be able to produce special mixes if given the lead time to procure the required materials. Concrete from these plants is delivered to the job site in transit mixers or batch trucks with capacities from 6 CY up. Concrete is sold by the delivered cubic yard. The contractor must provide all personnel for placement of the delivered concrete.

Concrete requirements are determined by calculating the volumes of the structures to be constructed of concrete. A waste factor of about 5% should be added to account for waste due to spills or delivery of concrete in standard batch sizes which exceed the needs of the current pour. When pouring against surfaces which are impossible to fine grade (such as rock excavation or trench walls), allow 3 to 12 in. outside the payline (edge of pour as measured from drawing). The volume taken up by the resteel is usually ignored in calculations except on extremely dense resteel situations such as nuclear containments or missile silos.

Concrete Placement

Placement of the concrete is a labor and equipment cost item. It can be a major or a minor cost item depending on the circumstances. The placement costs encompass all actions required to move the concrete from the delivering batch truck or mixer to the point of placement. It also includes use of vibrators to consolidate the concrete in the forms.

Direct Placement or Chute Placement.   The least costly method of concrete placement is realized when the concrete can be placed directly from the chute of the delivering batch truck.

Concrete Buggies.   Two wheel hand-pushed buggies, motorized buggies, or wheelbarrows may be used to move concrete from the delivery source to point of placement. This introduces an additional labor and equipment cost.

Concrete Pumps.   Skid-, trailer-, and truck-mounted pumps are now available for transport of concrete horizontally and vertically from the point of delivery.   Typical concrete delivery pipe has an inside diameter of 4 in. or more and comes in sections which can be assembled to reach remote pour locations.   Some pump trucks and trailers are equipped with booms which can reach outward or upward in excess of 100 ft. Pumped concrete is used for significant pours where continuous flow of concrete can be maintained since the concrete cannot be allowed to stand in the lines. For intermittent deliveries to remote locations, a conveyor is better suited. Pumps are available which deliver more than 100 CY/hour over distances exceeding 500 ft.

Concrete Conveyors.   Conveyors may be pedestal-, trailer-, or wheel-mounted.   These are standard conveyer belts adapted for concrete transport and may be assembled in tandem to provide delivery of concrete to any point. They may be angled upward or downward up to 30 degrees. Turret mounted versions are available which rotate and extend to provide delivery over a wide area. Conveyors can deliver up to 150 CY/hour.

Crane and Bucket. An older, but still very common, form of concrete delivery is the crane and bucket. Concrete is delivered into a crane-suspended bucket of 1 or more cubic yards capacity. This bucket is hoisted by the crane to the point of delivery. The bottom of the bucket has a gate which workers at the point of delivery can activate to release the load. Delivery may be to the point of placement or into concrete buggies or onto conveyors which deliver it where needed.

Concrete Elevators. On high-rise construction, a contractor may choose to use an elevator mounted against the side of the structure which transports a specially designed hopper. The hopper is filled at ground level, transported vertically by the elevator, and the contents discharged into waiting concrete buggies at the level where concrete is being placed.

Tremies. A tremie is a special form of pipe used in conjunction with any of the methods of concrete delivery to place concrete underwater. A typical tremie is a pipe about the diameter of a basketball and is positioned vertically in the volume to be filled and about one diameter's distance off the bottom (Fig. 13-10). A plug or "pig" is placed in the pipe and concrete is poured over it forcing the plug to move down the pipe. This plug keeps water from mixing with the fresh concrete as it moves down the pipe. Eventually the concrete forces the plug out the end of the pipe and it floats to the surface and the concrete flows out

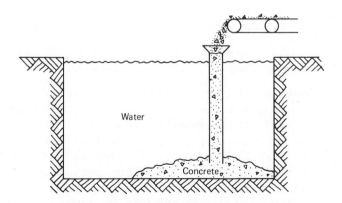

Fig. 13-10   Using a tremie to place concrete under water.

over the bottom. Thereafter, the end of the tremie is kept in the fresh concrete to maintain the seal but it will be moved up as more concrete is added. A tremie with a hopper on the top is used when concrete is delivered by other than pump. Tremies may be hooked directly into pump lines.

Vibration of Concrete. Concrete is normally subjected to vibration as a means to move the concrete around the resteel and eliminate voids. Handheld vibrators which are shaped like long pipes with closed ends are engine driven or pneumatically operated. The handheld vibrator is well suited for slabs or any pours where the workers can reach down into the concrete with the vibrator. Another form of vibrator occasionally used is the form vibrator. It is attached to the outside of forms creating vibration through the forms. However, form vibrators can easily cause mix segregation so their use is generally restricted to special applications.

Work Package Collection Sheet for Concrete Placement

Figure 13-11 is another example of a Work Package Collection Sheet, this time for concrete placement in a slab using direct chute technique. In the Notes section the duration of the project was determined to be 6 hours, but this was rounded up to 8 hours since it is difficult to deal with work crews on less than a half day basis, and, on a relatively small pour like this, there will be time lost in getting organized and shutting down. Productivity data comes from company historical files, cost information from current data files. This package does include finishing of the slab since slabs are finished as soon as placed and screeded. If this had been a wall, finishing would await removal of forms and would be handled as a separate work package.

Concrete Finishing

Some finishing operations are performed on the fresh concrete while others are completed after the concrete has set.

Slab Finishing. A fresh slab pour is first *screeded*. This is a basic leveling operation to bring the surface of the concrete to grade. It is done by using a long wooden board or metal bar that spans the distance between the edge forms. The screed is moved back and forth over the surface of the concrete keeping fresh concrete ahead of the screed so that low spots can be filled. The slab is then *darbyed*. A darby is a flat piece of wood or metal, 3 or 4 feet in length, that is moved over the surface with a handle to smooth and level the surface. If the concrete is to be *edged*, this operation can begin after the concrete begins to set.

WORK PACKAGE COLLECTION SHEET

System/Structure Identifier 02 • 133 | Crew Level Work Package Identifier 03 • 131 | Description CONCRETE PLACEMENT, FLOAT FINISH GROUND SLAB, BUILDING 2

Permanent Materials (PM)

Resource Code	Description	Unit	QTY	Unit Cost Low	Unit Cost Target	Unit Cost High	Extension
1325	CONCRETE, 2500 PSI	CY	135*		30.90		4171.50
	*5% WASTE INCLUDED						
							4171.50

Other Materials and Supplies (M&S)

Installed Equipment (IE)

Crew Labor (L)

Resource Code	Description	NR.	Cost/Hour Low	Cost/Hour Target	Cost/Hour High	Extension
0291	FOREMAN	1		10.90		87.20
0293	LABORERS	4		10.40		332.80
0292	FINISHER	1		12.85		411.20
						831.20

Equipment Not Charged As Job Indirects

Resource Code	Description	NR.	AVAILABILITY Periods	AVAILABILITY $/Period	USE Hours	USE $/Hour	Extension
7311	VIBRATOR, GED	2 @			8	3.25	52.00
7319	HAND TOOLS	1 @			8	1.00	8.00
7312	FINISHER, GED	1 @			8	4.50	36.00
							96.00

Productivity

Base Unit for Productivity CY CONCRETE

Total Quantity Base Unit 128 CY

Duration (Crew-Hours)	Low	Target	High
		8	

Escalation Rates (%)

Materials

Labor

Equipment

Notes

$$DURATION = \frac{128\,CY}{22\,CY/HR} = 6\ HRS$$

ALLOW 8 HRS WITH STARTUP AND CLEANUP

Cost Summary:

PM =	4171.50
M&S =	
IE =	
L =	831.20
CE =	96.00
TOTAL =	5098.70

Fig. 13-11 Work Package Collection Sheet for concrete placement.

A handheld metal tool is used for this edging which gives the edges a rounded cross-section. If the slab is to receive control joints, this is done using a *jointer* which is another handheld tool that is basically flat-bottomed but contains a ridge down its center for the full length of the tool. This tool indents the concrete as between sidewalk slabs. The slab is then *floated* using either hand or powered floats. A float is a flat-blade-like tool that is worked over the surface of the concrete to insure embedment of particles, remove small imperfections, and to consolidate mortar at the surface. The surface is then *troweled*, a process similar to floating, to give the final smooth surface.

There are a number of special finishes that can be applied. A broom may be drawn across the surface to provide a *broom finish* which gives a nonslip surface. *Exposed aggregate* is another popular finish. For this, the mix is processed through the float stage as described above. Then, by means of a broom and a jet of water from a garden hose, the mortar at the surface is removed to expose the aggregate.

Wall Finishing. Unless walls and columns have been precast and lifted into place, their finishing normally awaits form removal. Form removal will expose form ties and some surface voids. Basic wall finishing involves removal of ties and patching of voids with grout. The surface may be *burlap rubbed* with grout to give a grain texture overall. However, this method has been largely replaced by brush-applied epoxy finishes. A *bush hammer* finish roughens the surface by creating many small pits. In some cases, special finish treatments are imparted to the concrete surfaces during placement by using form liners. Even with these, some touchup finishing may still be required following form removal.

Curing

Concrete gains strength with time in the presence of water. Curing involves the maintenance of water in the fresh concrete for about 7 days. Optimally, curing protection should start as soon as the darby operation is complete on slabs. Damp burlap covers or moisture barriers should be placed over the slab and removed only temporarily for floating and troweling. As soon as the concrete has its initial set, it should be maintained in a wet condition. Direct sprinkling of the slab is satisfactory only if done often enough so that the slab does not dry out between sprinklings. Sprayed sealing compounds are available which seal the surface of the concrete to prevent moisture loss in the event wet curing is not possible. Keeping walls and columns in the forms retards moisture loss. Curing should be treated as a work package even though no labor or materials are expended because it does require time and must be treated as an activity on a network schedule. Contract specifications may dictate exact curing methods to be used. Basically, the obligation of the contractor is to insure that the concrete produced is of the quality specified through proper mix selection and curing.

Waterproofing and Insulation

Technically, waterproofing and insulation of concrete is not a part of the concrete operation but certain waterproofing and insulation must be installed prior to or after pouring concrete and is mentioned at this time. One of the most common waterproofing materials is a sheet of polyethylene or fiber-reinforced waterproof paper placed under ground slabs to prevent moisture movement through the slab. Water-stops are also installed prior to the concrete pour. These are copper, rubber, or synthetic material strips or other shapes embedded in the concrete at construction joints to prevent moisture movement through the joint.

Other types of waterproofing may be applied to completed concrete surfaces. Bituminous waterproofing is normally painted on the exterior surface of basement walls before backfilling or on any concrete surface which requires such waterproofing.

Special types of insulation may be installed around the perimeter of a slab to prevent building heat loss. This insulation typically comes in sheets and is waterproof. A typical design calls for 1 to 2-in. thick sheet extending 24 in. below grade level of the slab. It would be placed inside the slab forms prior to pour.

CARPENTRY

Carpentry is divided into rough carpentry and finish carpentry. Rough carpentry includes basic wood framing and sheathing of structures plus form construction, while finish carpentry includes door and window framing, cabinet work, paneling, and other finish work. A typical small general contractor will accomplish rough carpentry tasks with force-account labor while subcontracting at least part of the finish work. Larger contractors will maintain cabinet shops and do their own finish work as well.

Wood Materials

Dimension Lumber. This term applies to those pieces of lumber cut directly from a log. Individual pieces may be sold with their original sawed finish or one or more surfaces may be dressed (planed).

TABLE 13-3. Dimension Lumber

Nominal Dimension, (inches)	Dressed Dimension, (inches)
1	$\frac{3}{4}$
2	$1\frac{1}{2}$
3	$2\frac{1}{2}$
4	$3\frac{1}{2}$
5	$4\frac{1}{2}$
6	$5\frac{1}{2}$
8	$7\frac{1}{4}$
10	$9\frac{1}{4}$
12	$11\frac{1}{4}$

Whether dressed or not, lumber is sold by the original sawed dimensions (nominal size), not the dressed size. Since dressing reduces the dimensions of the lumber, this reduction must be accounted for in the measurement process. Table 13-3 shows the relationship between nominal size and dressed dimensions. Thus, a dressed 2 × 4 has actual dimensions of $1\frac{1}{2}''$ by $3\frac{1}{2}''$.

Dimension lumber comes in 2-foot increments of length and widths shown in Table 13-3. It is often difficult to find lengths over 16 ft and specialty woods may be available only in random lengths and widths.

Dimension lumber is ordered on the basis of the foot-board-measure (FBM). By definition FBM is

$$\frac{\text{Nominal width in inches} \times \text{Nominal thickness in inches} \times \text{Length in feet}}{12}$$

Another way of looking at the FBM is to think of it as 144 cubic inches of wood in the undressed state, or as a piece of undressed lumber 1 in. × 12 in. and 1 foot long. In all cases, use the undressed dimensions for calculating FBM.

Dimension lumber is graded as to its suitability for finish or utility purposes. Grading systems vary with intended function of the lumber product (framing, decking, scaffolding, etc.) but common grades are

1. Select Quality, Suitable for natural finish
   Grade A: Practically clear
   Grade B: Generally clear

2. Select Quality, Suitable for paint finish
   Grade C: Suitable for high quality paint finish
   Grade D: Better than common and suitable for painting
3. Common Quality, Suitable for use without waste
   #1: Sound and tight-knotted
   #2: Similar to #1 but not quite as high quality
4. Common Quality, Suitable for use with waste
   #3: Some features leading to waste
   #4: Low quality

Lumber is sold by the FBM or MFBM (thousand board feet). Prices depend upon the type of wood, the dimensions, and the grade. Normally, an estimator will determine the total foot-board-measure requirements of each size and grade of dimension lumber required and request quotations from suppliers for use in the bid.

Plywood. Plywood is used for sheathing, subflooring, and cabinet work. The standard plywood size is 4 ft × 8 ft which is both nominal and actual size. Plywood is available in thicknesses from $\frac{1}{4}$ in. to 1 in. Plywood is ordered by area (SF) or by sheets. In addition to specifying the area requirements, the following must be part of the specification:

1. *Quality.* The exterior plies are graded from A to D. A is clear and suitable for natural finish while D may have unplugged knots. The two exterior plies may be of different grade; in fact that is most common. For example, C-D is usually specified for wall and roof sheathing and subflooring. A and B grades are reserved for cabinet work. Inner plies on plywood are normally grade D, occasionally C.

2. *Exterior or Interior.* Waterproof glues are used on exterior plywood and on interior plywood to be used in areas exposed to moisture. It is essential that all plywood that may be exposed to direct moisture or dampness be of exterior grade to avoid ply warping or separating.

3. *Group.* The various woods are placed in one of five groups based on strength with Group 1 high and Group 5 low. The Group classification in combination with the plywood thickness determine the spacing of joists for plywood decking and subflooring. For example, a Group 1 plywood that is $\frac{3}{8}$-in. thick is structurally equivalent to a Group 2 plywood that is $\frac{1}{2}$-in. thick (24-in. joist spacing for roof decks). On some plywood, the Group number is shown on the grading label. On others, a recommended spacing of joists is given which is equivalent to the Group number.

4. *Thickness.* Actual and nominal thicknesses are the same for plywood.

Interior Paneling.   Many designs of interior paneling are available for wall finishes.   The surface veneer may be a natural wood, vinyl, formica, or other bonded surface.   Thicknesses vary but are usually about $\frac{1}{8}$ in.   Sheets of interior paneling are 4 ft × 8 ft and are ordered by area or sheets and the specific style desired.   Other dimensions may be specially ordered but will be significantly higher in price per square foot.

Exterior Siding.   Exterior siding may be finish grade dimension lumber which has been shaped for use as siding.   It often has a tongue and groove design for vertical application or is shaped for use as horizontal lapped siding.   Plywood siding is also available with exposed plies of many designs.   Most plywood siding comes in the standard 4 ft × 8 ft sheet but some patterns are available in widths suitable for lap siding and in lengths up to 16 ft.   Exterior siding is ordered by area and by style.

Determining Requirements

Carpentry material requirements must be counted or measured from the drawings.   With so many different dimensions and grades of wood products, it is essential that the estimator keep them segregated since costs vary considerably with size and grade.   Additionally, the estimator must remember to include an allowance for nails and other hardware incorporated into the carpentry work.   As previously mentioned, dimension lumber has cross-sectional dimensions less than nominal and this must be accounted for in some calculations.   Finally, there is always waste in the cutting of lumber products, some lumber is too warped to be fully usable, and some gets damaged or stolen at the job site.   Thus, quantities determined from quantity takeoff should be increased about 10% in the case of plywood and at least that much for dimension lumber to account for these problems.

Some lumber dealers now provide prefabrication service for trusses or other assemblies.   If this service is utilized, a contractor will not have to get involved in shop drawings, jig construction, and the fabrication of these complicated assemblies.

Carpenter productivity is normally measured in the same units as the materials with which they work.   For example, framing productivity is in terms of FBM per employee-hour; plywood is SF per employee-hour.

MASONRY

Masonry includes concrete block, brick, clay tile, and stone masonry. Masonry requirements are calculated using overall exposed surface area and number of layers (wythes).   For block, brick and tile, this is translated into numbers of units required. For stone, quantities are expressed as square feet of exposed area, tons, or cubic yards depending upon the material.

Masonry walls are solid, hollow, veneered, or reinforced. Solid walls have all joints filled with mortar. Hollow walls have a cavity left between the inner and the outer wall. For a reinforced wall, resteel and grout are introduced between inner and outer walls or in the void spaces of the block. A veneer wall is placed over the exterior sheathing of a framed structure.

The cost of masonry work includes the block, brick, tile, or stone, the mortar, certain accessories, the labor for installation, plus charges for the mortar mixer and hand tools. On large jobs where materials must be moved horizontally or vertically for some distance there will be additional materials handling equipment and labor charges.

Concrete Block

Concrete block is produced from sand aggregate, cinders, or special aggregates. Block is used for load bearing walls, curtain walls, chimneys, and decorative walls. There are many designs available, but the most common is the hollow core block with nominal face dimensions of 8 in. × 16 in. and thicknesses of 4, 6, 8, 10, or 12 inches. Actual dimensions are $\frac{3}{8}$ in. less each way. Thus, if a $\frac{3}{8}$-in. mortar layer is used, a single 8-in. × 16-in. block will cover an area 8 in. × 16 in. while any other thickness of mortar will provide different coverage.

Another wall forming technique is to stack concrete block without any mortar between joints. Then, a fiberglass-reinforced mortar is used to apply a plaster-like coat to the exposed surfaces of the block. The resultant wall is claimed to be at least equivalent to a conventional block and mortar wall, if not stronger. For this type of wall, actual block sizes must be used when determining requirements.

Reinforced concrete components may be constructed with block by inserting reinforcing bars in the void spaces of the block and filling the voids with mortar. Using this procedure, the equivalent of reinforced concrete columns and beams can be built into a structure without the usual forming problems.

Brick

Brick is a clay masonry unit. It may be solid or have several core holes. Nominal sizes of typical bricks are as listed in Table 13-4. Actual sizes are approximately $\frac{3}{8}$ in. less but different clay shrinkage factors during baking produce some size variation. Therefore, the estimator should use nominal dimensions and assume they include a mortar thickness that a mason will adjust to accommodate variations within a given style of brick. Bricks are normally ordered by the thousand (M).

TABLE 13-4. Brick Types and Dimensions

| | Nominal Dimensions | | |
| Brick Name | Thickness | Height | Length |
	(all dimensions in inches)		
Modular	4	$2\frac{2}{3}$	8
Engineer	4	$3\frac{1}{5}$	8
Economy	4	4	8
Double	4	$5\frac{1}{3}$	8
Roman	4	2	12
Norman	4	$2\frac{2}{3}$	12
Norwegian	4	$3\frac{1}{5}$	12
King Norman	4	4	12
Triple	4	$5\frac{1}{3}$	12
SCR	6	$2\frac{2}{3}$	12

Clay Tile

Clay tile is the hollow form of clay brick. It may be used for structural backing or facing. There are many designs used, none being particularly dominant. Like concrete block, their nominal size exceeds the actual size to allow for the mortar.

Stone

The many variations in type and size of stone require that an estimator make a special study of any stone masonry work to determine methods of ordering, surface area yield, and mortar requirements.

Mortar

Mortar for masonry work is produced on the job site using portland cement, hydrated lime or lime putty, masonry cement, sand, and water. A number of different standard mixtures have been developed to meet various masonry requirements so quantities of ingredients are definitely a function of the specified mortar. Masons utilize specially designed portable mortar/plaster mixers of 2 or more CF in size to mix the mortar.

Other Requirements

Various ties and other reinforcement are used to tie bricks to sheathing or to strengthen multiple wythes of masonry. These may not be shown on the drawings but are installed as a matter of good masonry practice. Sheet metal flashing may be specified on the drawings at various points for waterproofing. Lintels over openings, expansion joints, specially cast gaps, and various embedments may also be part of the wall design.

Collection Sheet Items for Placement of Masonry

Following is a checklist of potential resources of construction (permanent materials, materials and supplies, installed equipment, craftsmen, and construction equipment) that an estimator should consider for inclusion on the Work Package Collection Sheet for placement of masonry. Similar checklists will be provided in subsequent paragraphs for other work tasks.

Permanent Materials:
    Brick, block, or tile
    Mortar
    Ties
    Flashing
    Reinforcing bars (if used)
    Concrete to backfill around reinforcing (if used)
    Lintels and caps
    Expansion joints, other embedments
Materials and Supplies:
    None
Craftsmen:
    Masons and helpers
    Sheetmetal worker (for flashing, if required)
Construction Equipment:
    Mortar mixer
    Forklifts, hods or other equipment for masonry transport
    Scaffolding
    Mason hand tools
    Sheetmetal worker hand tools (if required)
Method of Estimating:
    Determine area requirements of finished wall. Area coverage of one brick, block, or tile is equal to *actual* length plus one mortar thickness multiplied by height plus one mortar thickness. Divide area requirements by coverage per unit to get total

requirement. Adjust as necessary for unusual placement patterns. Mortar requirements will vary from about 13 to 18 CF per thousand bricks, depending on brick size and mortar thickness; for concrete block more is required but this figure also is dependent on block and mortar thickness so should be calculated. If masonry is delivered and handled on pallets, a waste allowance of 1 to 2% should be adequate; loose handling creates more breakage and waste, particularly with concrete block.

Productivity Considerations:

Productivity is highest when runs are long and continuous. If there are many openings, short runs, and pattern changes, productivity will suffer.

STRUCTURAL STEEL

Installation Procedures

Structural steel members include standard rolled steel shapes (I-beams, wide-flange beams, angles, etc.), and special steel members built up from standard rolled steel shapes (plate girders, open-web trusses, etc.). Structural steel is used for framing of many buildings and for other applications where high tensile and compressive strength with a relatively small cross-section is required.

When erecting structural steel, columns are initially set onto foundations in which anchor plates and bolts have been set. These are followed by temporary placement of horizontal and diagonal members. Following final plumbing and alignment of key members, all members are permanently connected. The process is repeated vertically until all members are in place. Some members may be left out temporarily to permit lifting and setting of large equipment, ductwork, or other bulky components at a later date. Alternatively, these items may be temporarily or permanently placed as each level is reached. Bolting or welding of one side of connections while steel is on the ground or prefabrication of subassemblies composed of several steel members may speed production.

Most structural steel connections are completed with high-strength bolts although some structures are designed with some or all arc welded connections. Steelworkers erect the steel and install temporary bolts or connections during the basic assembly process. They follow up later to install and tighten the final bolts. If welded connections are used, welders do the welding portion. Nondestructive testing (radiographs, liquid penetrant dye, ultrasonic, or magnetic particle) may be specified.

Collection Sheet Items for Erection and Temporary
Steel Connection

Permanent Materials:

Structural steel members by American Institute of Steel Construction (AISC) designation and grade

Angles, seats, gusset plates, other connections

Other Materials and Supplies:

Temporary bolts or other temporary fastening devices

Craftsmen:

Steelworkers

Hoist operators

(Survey support if needed)

Construction Equipment:

Hoisting equipment

Steelworker hand tools

(Survey equipment if needed)

Method of Estimating:

Most structural steel is ordered through a fabricator who will develop shop drawings, cut, and punch the steel shapes and matching connections, attach some connections, and paint all exposed surfaces. Fabricators provide a lump-sum bid for this purpose. Add any costs of transportation and handling, plus bolts, labor, and construction equipment. Allow about 10% waste for bolts.

Productivity Considerations:

Productivity is affected by size of members, number of hoists required to lift the member, elevation of work, ease of access to final position, and number of delays during erection sequence for positioning of bulky equipment or other components. Allow extra time for plumbing and alignment of key components.

Collection Sheet Items for Bolting of Structural Steel

Permanent Materials:

High-strength bolts

Other Materials and Supplies:

None

Craftsmen:

Steelworkers (in pairs or with helper)

Air compressor operator

Construction Equipment:

  Impact wrenches

  Air compressor

  Steelworker hand tools and rigging equipment

  Safety nets

  Scaffolding and work platforms

Method of Estimating:

  Determine number of bolts by counting or factoring from weight of steel. Add 5% for waste. Bolts ordered from steel supplier by specification, size, and length.

Productivity Considerations:

  Productivity will be a function of steelworkers access to bolt holes and the accuracy of hole alignment during erection. Productivity is also affected by the number of bolts that can be installed from one position and ease of worker movement to subsequent positions.

Collection Sheet Items for Field Welding
of Structural Steel

Permanent Materials:

  None

Other Materials and Supplies:

  Welding rod

  Nondestructive testing supplies (if not otherwise accounted for)

Craftsmen:

  Welders and helpers

  Nondestructive weld testers (if not otherwise accounted for)

Construction Equipment:

  Welding machine and welding cables

  Welder personal protective equipment and hand tools

  Grinders for joint preparation (if required)

  Nondestructive test equipment

Method of Estimating:.

  Welding productivity usually expressed as linear feet per day for given type of weld. Welding procedure specifications governed by American Welding Society and described in the AISC *Manual of Steel Construction*. Identify number and length of welds by AWS specification as basis for total employee-hour and welding rod requirements. As much welding as possible should be done prior to steel erection. Thus, it is desirable to break welding operation into welding on ground and aerial welding.

Productivity Considerations:

Some welds require more than one pass to complete. Productivity is also affected by ease of access to weld and weld position (flat, vertical, overhead). All edges to be joined must be clean and prepared in accordance with AWS specifications. If surfaces have not been prepared by fabricator, allowance must be made for shaping of edges with grinders prior to erection.

PRESSURE VESSEL PIPING

Installation Procedure

Extensive pressure vessel piping is associated with steam-electric generating plants, process plants, and other industrial installations (boilers, steam piping, reactor piping, etc.). Connections of pressure vessel piping are arc welded using either shielded metal arc welding (SMAW or stick welding), tungsten-inert gas (TIG) welding, metal inert gas (MIG) welding, or submerged arc welding. The latter two processes are used extensively in shop welding since both can be automated. Stick, TIG, and some MIG welding are used in the field. Oxyacetylene welding is never used for pressure vessel welding although this equipment may be used for steel cutting operations.

Pressure pipe/vessel welding is more rigidly controlled than structural welding. Whereas structural welding is governed by American Welding Society standards, pressure vessel welding is governed by American Society of Mechanical Engineers (ASME) codes. Procedures for each type of weld are established by written Welding Procedure Specifications that have been precertified. All welders must be tested and qualified in each type of pressure vessel weld they perform. In addition, all pressure vessel welds must be visually inspected by an independent authorized inspector during and after the welding operation. Preheating and/or postwelding stress relieving may be specified for a weld. Nondestructive testing (radiographic, liquid dye, ultrasonic, or magnetic particle) is also required. For welding operations subject to nuclear quality assurance, the inspection and recording function is particularly rigid.

Joint preparation and fitting of pipe for welding is a major operation. All edges to be joined must be shaped in strict accordance with specifications. Figure 13-12 shows a typical joint preparation detail. Normally, pipe sections are cut and bent by commercial fabricators and joint preparation is included. Pipe sections must be fitted (aligned) for

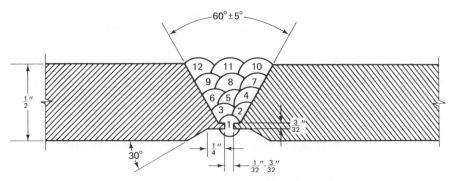

Fig. 13-12 Detail showing joint preparation and number of passes for pressure pipe weld.

welding. Since pressure pipe is very thick compared to pipe used for ordinary plumbing installations, it is very heavy and many special clamps and other fitting hardware will be required to bring and hold the pipe sections in alignment. When the sections are properly fitted, the welders will complete the weld. A typical pressure vessel weld consists of many passes, as shown in Fig. 13-12, so weld productivity is not based on weld length as was the case with structural welding. Instead, productivity is normally expressed in terms of employee-hours per weld of a certain type. Some welds in nuclear plants require several hundred employee-hours each.

A contractor will normally prefabricate as many sections of pipe as possible in a pipe shop on site since control is much easier and productivity much higher than for field welding. Pipefitting and welding of pipe in final position can be very difficult and time consuming since strict tolerances are established for fitting. Various hangars or other supports may be installed as pipe is fitted.

Following installation of a line or system there may be a requirement for cleaning, blowing, and testing. Treat this as a separate work package.

Collection Sheet Items for Pipe Fitting

Permanent Materials:
    Pipe sections
    Hangars or other support items
Materials and Supplies:
    None
Craftsmen:
    Pipefitters
    Hoisting equipment operators (for field welding)

Construction Equipment:

  Hoisting equipment

  Joint preparation tools (if not shaped by fabricator)

  Rigging accessories

  Pipefitting clamps

  Pipefitter hand tools

Method of Estimating:

  Pipe sections are ordered through commercial fabricators who will prepare shop drawings, cut and bend pipe, prepare joints, and prefabricate into subassemblies for shipping. A lump-sum bid for the service is possible. Add costs of transportation and handling, plus site labor and equipment for installation.

Productivity Considerations:

  Take maximum advantage of on-site prefabrication in shops. Field pipefitting productivity subject to congestion in area of installation, number of joints in the subassembly being fitted, and accuracy of fitting of previously installed components to which subassembly is to be fitted.

Collection Sheet Items for Welding of Pressure Piping

Permanent Materials:

  Backing strips and consumable welding inserts

Materials and Supplies:

  Welding rod (stick and some TIG) or wire (MIG)

  Inert shielding gas (TIG and MIG)

  Cooling water or air (TIG)

  Nondestructive testing supplies (if not handled separately)

Craftsmen:

  Welders and helpers

  Quality inspectors (if not handled separately)

Construction Equipment:

  Welding machines with cables and tubing

  Welder hand tools and protective equipment

  Nondestructive test equipment (if not handled separately)

  Preheating and stress-relieving equipment

Method of Estimating:

  Each type of weld should have an associated number of employee-hours. Summarize types of welds and extend to get total employee-hours required. Material requirements are handled similarly.

Productivity Considerations:

As with pipefitting, shop welding is more productive than field welding.

OTHER PIPING

Installation Procedure

Building Plumbing. This category includes piping associated with building water and wastewater systems for personnel. Plumbing work of this type must be scheduled by phases: underground connections, setting of piping which will be embedded in foundations, other interior rough plumbing, and finish work (final connections and fixtures). This work is most frequently accomplished by subcontract with a plumbing contractor on a lump-sum basis to include materials and installation so will not be further discussed in this chapter.

Other Piping. In industrial construction work a great variety of other piping systems may be included. Cooling water loops for power plants, fire protection systems, fuel oil systems, and compressed air lines are in this group. Although these lines do operate under pressure, they are not high pressure pipes in the sense that they are governed by ASME codes. In fact, most of this piping has threaded or flanged connections, instead of welded joints. Installation of this piping is subject to local and state codes. This piping may be underground, embedded in concrete foundations, hung within a structure, or placed on other above ground supporting framework. Many large contractors do this type of work with their own personnel.

Underground piping installation includes ditching, possible shoring and dewatering, placement of select fill or concrete bed on which pipe will rest, possible select backfill, and possible construction of thrust blocks at points of pipe direction change. Cleaning and hydrostatic testing usually follow completion of connections but are done before backfill. Piping to be embedded in concrete must be installed prior to resteel placement and any embedded joints thoroughly tested before concrete is placed. Installation of other piping must await installation or erection of required hangers or other supports.

Very large pipe sections are often ordered from a commercial fabricator who prepares shops drawings, cuts and bends pipe, attaches flanges and otherwise custom prepares the pipe sections. As with pressure piping, there is a limit to the size of pipe section (spool) which can be transported so additional prefabrication in on-site pipe fabrication shops is also common. On-site fabrication of all pipe sections utilizing pipe sizes that are 2 in. or less is normal on union projects.

The following collection sheet information is general and intended to cover common pipe installation situations. It does not include resources used for excavation or backfill of ditches in which pipe is placed or erection of hangars or other supports.

Collection Sheet Items for Piping Installation

Permanent Materials:
  Pipe
  Couplings, tees, valves, gauges, and other accessories
  Special supports or attachment hardware
  Flange bolts (if flanged)
  Gaskets or other sealants
  Thrust block materials for underground installation
  Select bedding material for underground installation
  Coating materials
Materials and Supplies:
  Fabrication expendables
  Shoring for ditches (if required)
  Gas, oil, or water for hydrostatic testing
  Materials for flushing and cleaning
Installed Equipment:
  Motor driven valves
  Other attached instrumentation
Craftsmen:
  Plumbers and helpers
  Equipment operators (hoisting and dewatering equipment, if used)
  Carpenters (for any forming or scaffolding)
  Ironworkers (for reinforcing)
Construction Equipment:
  Pipecutting and threading equipment
  Craftsmen's hand tools
  Hoisting and rigging equipment (if applicable)
  Pipe coating equipment
  Hydrostatic testing equipment
  Equipment for flushing and cleaning
  Scaffolding (for some overhead installations)
  Dewatering equipment (if required)

Method of Estimating:

Lump-sum bids for commercially fabricated items are adjusted for any transportation and handling costs. Other material costs are a function of quantities and unit costs at site. Employee-hours and equipment hours are subject to most variability.

Productivity Considerations:

Shop fabrication of components will generally speed productivity. Also partial assembly of components on ground adjacent to ditch or overhead support may speed operation.

ELECTRICAL INSTALLATION

Installation Procedures

Building Electrical Systems. Electrical service in a residence or commercial building to handle lighting, HVAC, and appliances is almost invariably handled by a specialty subcontractor who can be expected to bid the work on a lump-sum basis. Scheduling of this work is similar to the phases of plumbing: installation of embedded conduits prior to placement of concrete, rough wiring and placement of junction and out-let boxes, and finish work. Since this is a traditional subcontract item, it will not be discussed further here.

Industrial Electrical Systems. These electrical systems serve the usual building purposes plus provide electrical service to the many machinery, communications networks, and instrumentation components within an industrial facility. While some general contractors may choose to subcontract this work, the larger ones quite often handle this work with force-account resources.

There can be several phases to the electrical work. The first phase is the installation of a grounding and cathodic protection system, if one is required. This work must be coordinated with foundation work since all metal piling must be interconnected with grounding wire and other ground grids will be installed below or within foundations. The next phase is the erection of the basic power grid for the installation. Construction of substations, installation of transformers, and transmission lines are in this phase. Since portions of this system may be used for construction power, it is constructed early in the schedule. The third phase is the remainder of the electrical wiring and installation.

The transformers, motors, switch boxes, outlets, and other fixtures to be served electrically will be installed as separate work packages either before or after electrical wiring is in place. For most components, this is simply a rigging and setting operation. However, for some mechanical items, considerable time must be spent in balancing and alignment.

On union projects, electrical work has the potential for many jurisdictional disputes among electricians, linemen, pipefitters, steelworkers, and others. Mixed crews are often used to prevent conflict.

Because of the many wiring circuits in an industrial facility, an elaborate system for routing of wiring through ducts or on trays is often incorporated in the design. The ducts and trays are installed first, wire pulling schedules will depend upon this coding for proper routing of the wires. Concrete encasing underground duct banks is usually colored to identify it should later excavation be required. Wire trays the placement of these must be carefully coordinated with other work. Proper coding of ducts is essential during installation since very detailed wire pulling schedules will depend upon this coding for proper routing of the wires. Concrete encasing underground duct banks is usually colored to identify it should later excavation be required. Wire trays are suspended from ceilings, hung on walls, or supported on ladder-like frames built up from the floor. Each tray can handle many wires. For wire pulling, leader ropes or cables are threaded first. Then, one end of the leader is connected by a swivel to the wire mounted on a cable reel, the other to a pulling device. The wire is carefully pulled through the maze. A number of electricians may be needed during this operation to ensure that the wire does not bind at any point.

Termination of wires must be carefully controlled, particularly if wires are spliced directly to leads at units served. More modern designs of electrical systems maximize use of terminal boxes to ease this work. As with wire pulling, terminations are accomplished according to a strict schedule. Cadwelding may be used for very large conductor connections and terminations. Because of the special nature of these connections, they should be separately counted for estimating purposes (see the reinforcing steel discussion of cadwelding).

Collection Sheet Items for Underground Duct Installation

Permanent Materials:
    Duct by type
    Duct fittings and sealants by type
    Duct supports and spacers
    Concrete and resteel (for encasements or manholes)
    Prefabricated manholes (if used)
    Grounding cable (if specified)

Other Materials and Supplies:

    Forming materials

    Shoring (if used)

Craftsmen:

    Electricians or pipefitters and helpers

    Carpenters (for forming)

    Concrete workers (for concrete encasement)

    Equipment operators (for dewatering equipment, cranes)

Construction Equipment:

    Duct cutting and bending equipment

    Craftsmen's hand tools

    Dewatering equipment

Method of Estimating:

    Concrete work is estimated in usual fashion. Many types of ducts will probably be required and each has different cost and employee-hour requirements so these should be listed separately. Large field bends (over $2\frac{1}{2}$ in.) should be listed separately since they require more labor.

Productivity Considerations:

    Work will proceed much faster in areas where ducts are arranged in banks.

Collection Sheet Items for Cable Tray Installation

Permanent Materials:

    Cable trays

    Cable tray support materials

    Cable tray fittings and other hardware

Other Materials and Supplies:

    Tagging supplies

    Welding rod (if welding is used)

Craftsmen:

    Electricians or steelworkers and helpers

    Welders (if welded support structures are used)

Construction Equipment:

    Power tools for support fabrication

    Scaffolding

    Craftsmen's hand tools

    Welding machines (if used)

Method of Estimating:

Determine linear feet of trays and number of different types of supports (hangars, wall mounts, etc.) and factor costs and employee-hours from these.

Productivity Considerations:

Congestion of work area due to small space, interference with other systems, or other on-going construction operations are primary deterrents to reasonable productivity.

Collection Sheet Items for Wire Pulling

Permanent Materials:

Wire by type and size

Other Materials and Supplies:

Cable lubricant

Craftsmen:

Electricians and helpers

Construction Equipment:

Wire reel stands

Wire pulling equipment

Craftsmen's hand tools

Method of Estimating:

Factor employee-hour requirements from cable types and length.

Productivity Considerations:

Productivity will go down as length and diameter of wires increase.

Collection Sheet Items for Wire Terminations

Permanent Materials:

Wire terminals

Other Materials and Supplies:

Wire tags (identifiers or safety tags)

Cadweld expendables (if required)

Craftsmen:

Electricians

Construction Equipment:

Electrician hand tools

Electrical test equipment

Cadwelding equipment (if required)

Method of Estimating:

Employee-hours factored from number of terminations.

Productivity Considerations:

If properly coded terminal boxes are used, productivity will be higher than if all wires are individually spliced to leads from items serviced.

MECHANICAL INSTALLATION

Installation Procedures

Mechanical installation involves the delivery and placement of preengineered and prefabricated mechanical components in a structure. For a building, this includes systems such as HVAC, escalators, and elevators.   For industrial facilities it can include reactors, turbine-generators, pumps, conveyors, crushing units, gantry cranes, environmental control systems, and many other comparable items. In the construction of a typical residence or commercial building, the general contractor will subcontract for the procurement and installation of mechanical systems on a lump-sum basis so the estimating chore is relatively simple. On the other hand, an industrial contractor may handle all procurement, transport, and installation of mechanical components.

The estimation of costs for heavy mechanical components requires particular analysis by an estimating group. The basic procurement costs can probably be established by direct quotation. To this must be added transportation and installation. Transportation costs can be extreme if an item is oversize and overweight for routine commercial transportation. If so, special-purpose transporters must be available, bridges along routes of travel may have to be reinforced or bypasses constructed, power and communication lines may have to be taken down to permit passage of the item, and special roads constructed on site to support its movement. Hoisting of the item into final position may require renting of heavy hoisting equipment or construction of lifting superstructures. Setting of the equipment includes placement on previously installed foundations, and leveling and alignment including shimming and grouting. The leveling and alignment effort can be very significant on turbine-generator lines or large motor driven equipment. Balancing and testing will follow at a later time.

BASE COURSES AND PAVING

Construction Procedure

Roads, runways, parking lots, and other paved areas are constructed by placing progressively higher quality base course materials in layers over the natural subgrade material and topping this with a pavement such as asphaltic concrete or portland cement reinforced concrete.

Base course materials (crushed rock or other select fill) must usually be obtained from a source away from the project area. The cost of these materials includes a cost (royalty) per cubic yard or meter at the pit or quarry, processing cost, transportation to the site, placement, compaction, and grading.

Asphaltic concrete can be centrally produced or mixed on the job. The central-mix material is definitely of higher quality since mix control is much easier to assure. At a central-mix asphalt plant, heated asphalt is mixed with heated aggregate to produce the paving mix. This is then transported to the project site in dump trucks where it is spread with self-propelled paving machines over the base course which has been swept of dust and sprayed with a prime coat of hot asphalt by a bituminous distributor (tank truck with heater and spray bar). This is followed immediately by compaction with rollers, both smooth-wheel and pneumatic. A central-mix asphalt plant must be positioned relatively close to on-going paving operations so that mixes do not cool below minimum placement temperatures between production and placement. Tarpaulin covers are often used on the transport trucks to help reduce this heat loss. Central-mix asphalt plants are usually owned by the paving contractor although there are some plants in larger cities that operate similar to ready-mix concrete plants to serve small contractors. For the contractor who owns his or her own equipment, cost calculations must consider the same categories of cost associated with other owned equipment plus setup and movement costs. Great quantities of fuel are expended in heating asphalt and aggregate so estimators must be particularly careful in their estimating to insure that fuel costs are fully considered. Also, since asphalt is a petroleum product, it is a critical cost item.

Asphaltic concrete mixed-in-place is produced by spreading a layer of crushed aggregate, spraying this aggregate with hot asphalt from a bituminous distributor, blending the mix with a grader or rotary tiller, spreading the mix over the base, and compacting it. Often, a final seal coat of hot asphalt is sprayed over the surface and this may be followed with a blotter coat of finely crushed rock or sand.

Concrete pavements are usually reinforced at least with welded wire fabric, if not with deformed bars. Concrete is produced by commercial central-mix concrete plants or plants which the contractor owns and erects to serve the project. On road and runway projects, concrete is delivered to the placement site by agitator or mixer trucks, then spread, screeded, and often finished automatically with self-propelled equipment. On some road and runway placement operations, forms are used and the placement equipment rides these forms; however, many pavements and most curbs and gutters are now placed with machines which do not require forms. These machines are known as slipform pavers. On pavements for parking lots and driveways, the placement operation is comparable to that for a building ground slab.

On long, continuous reinforced concrete paving jobs such as runways and superhighways, the equipment is often arranged in a *paving train*. The lead equipment is a machine which planes the base course to close tolerances. This is followed by resteel placement equipment, the paver, finishing equipment, and a machine for spraying curing compound over the surface.

Base courses and paving operations are traditionally bid on a unit-price basis; that is, the bid is expressed in terms of a cost per cubic yard, square yard of finished surface, or similar unit. To obtain these unit costs the estimator must still treat each work package separately, determine the total costs for that package, and then use these costs as input for determination of the unit costs. This subject will be discussed more fully in Chap. 18.

Collection Sheet Items for Asphalt Pavement

Permanent Materials:
   Asphalt for asphaltic concrete
   Asphalt for prime coat
   Crushed rock
Other Materials and Supplies:
   Survey supplies (if not handled as job indirect)
   Heating fuel for asphalt and aggregate (other fuel in equipment cost)
Craftsmen:
   Central-mix plant operators
   Truck drivers
   Paving machine operators
   Laborers at paver
   Roller operators
   Bituminous distributor operator and assistant
   Survey personnel (if not handled as job indirect)
   Traffic control personnel
Construction Equipment:
   Central-mix plant
   Trucks
   Paving machines
   Steel wheel roller

Pneumatic roller

Bituminous distributor

Survey equipment (if not handled as job indirect)

Two-way radios (if not handled as job indirect)

Method of Estimating:

The cost of the central-mix plant must include not only the usual availability and use charges, but also the costs of erection, moving, and disassembly. Other than that, it is a matter of identifying all costs and spreading them over the total production. This type of paving is very sensitive to either wet or cold weather so adequate provision must be made for nonproductive days.

Productivity Considerations:

The key piece of equipment is the central-mix plant. Insure that enough trucks, pavers, rollers, and other equipment and personnel are available to keep plant operating at capacity.

SUBCONTRACT ITEMS

One of the characteristics of building construction contracting is the heavy use of subcontracts; some general building contractors may subcontract close to 100% of the work. Industrial and heavy/highway contractors tend to approach the other extreme and accomplish most of the work with force-account resources. Discussion thus far in this chapter has concentrated on representative work packages from both the building and industrial sector that a contractor will generally do with force-account resources since these are the ones for which detailed estimate preparation is appropriate. In the case of traditional subcontract work, an estimating section may still prepare estimates for purposes of evaluating and comparing subcontractor bids, but these estimates can be much more approximate and of the types discussed in Chap. 19. In other words, commercial handbook estimating is appropriate and there are a number of excellent handbooks on the market to facilitate this work. As far as the specialty contractor is concerned, he or she can be expected to have a close handle on costs since he or she does the same type of work over and over.

The following work items are listed as those typically subcontracted to a specialty contractor by most general contractors.

Architectural equipment installation (restaurant, laboratory, theater, etc.)

Curbs and gutters

Dewatering
Dry walls
Electrical (for buildings)
Elevators, escalators, and moving sidewalks
Flooring and carpets
Glass and glazing
Heating, ventilating, and air conditioning (HVAC)
Insulation for buildings
Landscaping
Piledriving
Plaster and stucco
Plumbing
Precast concrete
Rail spurs
Rigging (heavy)
Roofing
Security fencing and systems
Sprinkler systems
Waste water treatment
Wells

SUMMARY

There are two major objectives in the preparation of an estimate for purposes of bidding: first, to derive a total estimated cost; and second, to establish a basis for control of the project should the bid be successful. Accomplishment of these objectives is possible only through a detailed evaluation of resource requirements for each work package intended for force-account execution and reception of firm bids for subcontract work. Through a work packaging approach to estimating, these objectives can be met.

REVIEW

Exercises

1. Figure P13-1 represents a portion of an earthwork project. Existing contours are solid lines; proposed contour lines are dashed. Contour interval is 1 foot. Calculate cut and fill using:
    a. Grid method
    b. End area method

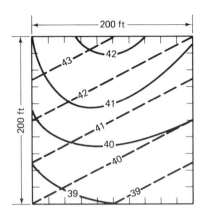

2. The swell factor for a soil is 20% and the shrinkage factor is 95%. If 10,000 cubic yards (bank) are to be cut, how many scraper loads will be required if each scraper can carry 20 CY heaped? How much fill will the cut material provide?

3. A concrete slab, 20 ft $\times$ 40 ft is to have #4 resteel reinforcement each way on 12-in. centers. Calculate the amount of resteel to order.

4. Determine the total FBM of the following list of dimension lumber:

Item	Quantity
2 in. $\times$ 4 in. $\times$ 8 ft	32
1 in. $\times$ 6 in. $\times$ 12 ft	18
1 in. $\times$ 8 in. $\times$ 10 ft	4
6 in. $\times$ 6 in. $\times$ 10 ft	8

5. How many square feet of brick veneer can 1000 Roman style brick be expected to yield?

6. Using $\frac{1}{2}$-in. mortar joints, how many concrete blocks are needed for 5000 SF of wall?

Questions in Review

1. Define shrinkage and swell factors.

2. Why is earthwork normally bid on a unit-price basis?

3. What cost items are included in developing a typical unit-cost figure for earthwork cut and fill work?

4. How can forming costs be reduced?

5. What costs are included in the total cost for placing resteel?

6. Why might a contractor set up his or her own concrete batch plant on the project?

7. Reinforcement on a drawing is identified as 6 $\times$ 6—4/4. What does this mean?

8. If you have both concrete pumps and concrete conveyors available for use, what factors will influence your choice of items for delivering concrete to a remote point on a project?

9. What is screeding of concrete?
10. How is exposed aggregate finish created on concrete?
11. Plywood for a structure is specified as "C-D, Group 3, Exterior." What does this mean?
12. Define foot-board-measure.
13. How is structural steel ordered?
14. What is the difference between nominal and actual sizes of concrete block?
15. What work packages are part of structural steel erection?
16. What differences are there between welding associated with structural steel and that associated with pressure pipe welding?
17. How does wiring for industrial circuits differ from that for a home?
18. What is a cadweld?
19. How do oversize/overweight mechanical components add to construction costs?
20. What difference, if any, is there between preparing Work Package Collection Sheets for unit-price bid items and those that are part of a lump-sum estimate?
21. Identify typical subcontract work packages.

Discussion Questions

1. Why do industrial contractors tend to subcontract less of their work than do building contractors?
2. On a large power plant, fossil or nuclear, there are thousands of welds involving different types of steel, different thicknesses of metal and different diameter joints. How would you propose to catalog estimating data to assist in the estimating of employee-hours and materials for these welding packages?

JOB INDIRECTS
AND MARKUP

14

INTRODUCTION

In this chapter we will complete our study of the individual components of cost that comprise the total bid price arrived at by a bidding contractor. These final components come under the headings of job overhead and indirect costs and markup. Each of these is significant in amount and so requires the same care in evaluation as the direct project costs.

JOB OVERHEAD AND INDIRECT COSTS

Job overhead and indirect costs (job indirects for short) encompass all those costs which can be attributed to a given job but which cannot be conveniently allocated to specific work packages. Many categories of cost can become part of this classification for a given job and every job, no matter how small, will contain some of these costs. Total job indirect costs depend on size and type of project and the variation is of such range that an estimator is best advised to determine job indirect costs by itemizing all anticipated costs, not by application of a flat percentage, unless the contracting operation is very repetitive and standard rates can be employed with little possibility of error.

Table 14-1 contains a listing representative of the many items which are potential job indirects. This particular listing is for a very large project and would be scaled down as necessary to fit a smaller project. The following sections highlight some of these items.

Personnel

Figure 14-1 is an organization chart for a typically large project to emphasize that staffing of such a project can include many functions.

TABLE 14-1. Job Indirects Checklist

Salaries, Supervision
Construction Manager
General Superintendent
Excavation Superintendent
Concrete Superintendent
Carpenter Superintendent
Rigging Superintendent
Welding Superintendent
Tunnel Superintendent
Electrician Superintendent
Equipment Superintendent
Quarry Superintendent
Chief Warehouseman

Salaries, Engineering
Chief Engineer
Office Engineer
Cost Engineer
Schedule Engineer
Materials Engineer
Draftsmen
Drawing Clerk
Field Engineers
Party Chiefs
Instrumentmen
Rodmen/Chainmen

Salaries/Wages, Other
QA Engineers
Safety Engineers
Mechanics
Plant Operators
Nurses
First Aid Workers
Secretaries and Clerks
Computer Operators
Training Personnel
Lab Technicians
Warehousemen
Guards
Janitors
Runners

Automotive
Automobiles
Station Wagons
Pickups
Ambulances
Tractor-Trailers
Cargo Trucks

Special-Purpose Trucks
Lubricators

Temporary Horizontal Construction
Access Roads
Construction Bridges
Drainage Structures
Rail Spurs
Laydown Areas
Fencing and Gates
Parking Areas
Environmental Protection

Buildings and Major Equipment
Project Office
Warehouses
Brass Alley
Change House
Carpenter Shop
Pipe Fabrication Shop
Welder Test Facility
Electrical Shop
Rigging Loft
First Aid Station
Tool Sheds
Powder House
Cap House
Paint Shop
Resteel Fabrication
Machine Shop
Equipment Maintenance Shop
Concrete Batch Plant
Quarry
Hoisting Equipment
Training Building
Visitor's Facilities

Support Systems
Water Supply
Compressed Air
Electrical
Site Communications
Inert Gas
Oxygen

General Expense
General Office Supplies
Office Furniture
Engineering Supplies
Engineering Equipment

TABLE 14-1. *Continued*

General Expense (Continued)	
Printing/Reproduction	Progress Photos
Computer Terminal	Sanitary Facilities
CPM Scheduling	Building Rental
Telephone/Telegraph/Radios	Move-In Costs
Utilities/Portable Toilets	Building and Grounds Maintenance
Signs	Exterior Lighting
Safety Equipment	Drinking Water and Ice
Permits and Licenses	Taxes
Advertising and Contributions	Bonds
Job Travel Expenses	Insurance
Testing and Laboratory	Payroll Burden Costs
Legal Fees	Backcharges from Others
Audit Fees	Consultants
Medical Supplies	Weather Protection

In the management group are the construction manager and superintendents, usually a separate superintendent for each major area or craft. On extremely large projects these individuals may have assistants. These are all salaried personnel and usually permanent members of a contractor's staff as compared to the foremen and craftsmen who are hourly wage employees and more temporary in their affiliation with the contractor. The project engineer may also have a large staff since his or her area of responsibility includes field design, cost control, scheduling, field estimating, review of contract drawings, preparation of shop drawings, field engineering (surveying), and operation of any field laboratories. The administrative staff can include a personnel office, payroll and accounting functions, warehousing, purchasing, and services. Additional job indirect personnel who may be either salaried or hourly wage include the employees in the maintenance shop and first aid station, those with the guard or building custodial forces, and operators of equipment listed under job indirects.

Automotive Expenses

As listed in Table 14-1 there are a number of vehicles needed for administrative support of a project. Some of these will be committed to a project almost continuously; others will be needed only for limited periods. Most of these vehicles are associated with the project office, the warehousing operation, or the equipment maintenance operation.

Temporary Horizontal Construction

To service the project a contractor may find it necessary to develop access and haul roads, build bridges and other drainage struc-

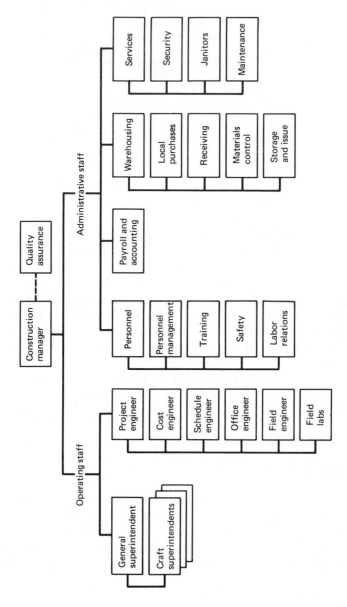

Fig. 14-1   Typical large project staff organization.

tures, build employee parking lots, develop laydown areas for equipment, construct temporary rail spurs, shape surface for drainage, install environmental protection structures such as sedimentation basins, and otherwise make the site usable by the construction force. In addition, for security and protection of the general public the contractor may find it necessary or be required to fence all or portions of the project.

Buildings and Plant

On a large project a contractor establishes a small city to support the work. The contractor must provide a brass alley for craft employees to check them in and out each day and possibly a change house where they can change clothes, eat lunch, or wait out inclement weather. A project office is necessary to serve the needs of the management and administrative staff. Depending upon the project, several types of warehouses may be needed. The most sophisticated warehouse is one with both humidity and temperature control for storage of sensitive equipment prior to installation. Heated and enclosed warehouses are the next category, with covered but unheated storage a third. Fenced and open laydown storage are final storage categories. A staff to operate the warehousing operation will be needed. That group is normally housed in one of the warehouses. A variety of fabrication and special-purpose structures as listed on Table 14-1 may be part of the project.

Construction equipment not accounted for in direct work packages is handled as a job indirect. General support cranes, batch plants, quarries and field generators are representative of equipment in this group. Remember that this equipment will have operating personnel who must be included with other job indirect personnel in the estimate.

Support systems are another category of plant that may be part of a project. Perhaps pumps and piping are installed to provide water or for drainage. Systems may also be installed to provide centralized distribution of compressed air, electricity, oxygen and inert gases for welding, and other operations. A large project will also have a complete phone network, loudspeaker system, and possibly alarm systems.

General Expenses

This category of job indirect costs is a catch-all for the many additional items that can be part of job costs. Referring to the listing of these in Table 14-1, many are self-explanatory. Several, however, require some emphasis and further explanation.

Insurance. Insurance may be required by the General and Special Conditions of the contract or may be purchased at the option of the contractor for his or her own protection. Although insurance is paid for by the contractor, its cost is passed on to the client as a job indirect or general overhead charge. Those policies purchased for a particular job would be treated as a job indirect. General coverage policies would be included with general overhead. Various types of insurance are detailed below.

Builder's Risk. Provides coverage against loss to structures or equipment by fire, collapse, earthquake, flood, other natural disasters, and vandalism. This type insurance may be purchased for individual jobs or a company may have a general coverage policy for all of its projects. An *equipment floater* extends coverage to construction equipment as well as structures. An *installation floater* provides protection when the contractor is handling other people's property or using their equipment. Other floaters are possible.

Public Liability. Provides coverage against bodily injury and property damage to third parties caused by the construction operations. (Note: Worker's Compensation provides bodily injury coverage for employees.)

Comprehensive Automobile Liability Insurance. Provides bodily injury and property damage coverage for vehicles under company jurisdiction. A *nonowned vehicle floater* is recommended. The cost of some of this insurance may be included within the ownership portion of the equipment cost where it is reflected in the periodic use charge and should not be duplicated under job indirects.

Special Policies. On certain projects special types of insurance are required or may be purchased. Examples are *steam boiler and machinery liability* and *nuclear energy liability* to cover special operations. Other policies available include those which protect a contractor from claims arising against his or her subcontractors or provide protection against theft of tools, equipment, and materials.

Bonds. As mentioned in Chap. 4, bonds are a form of insurance purchased by a contractor and issued by a surety which guarantees contractor performance to the client. There are three principal types of bonds.

Bid Bond. The bid bond guarantees that the contractor will undertake the work if he is the successful bidder. If required, it must be forwarded with the contractor's bid. It may be issued free by a surety to a good customer. In any event, its cost is nominal compared to the performance bond since it is for only 5 to 20% of the bid price and will be replaced by the performance bond if the contract is awarded. In some cases, the contractor is allowed to furnish a certified check, in lieu of the bid bond, which the client holds until contract award.

Performance Bond. This bond is often combined with the Payment Bond. It guarantees that the contractor will perform the contract to completion. It is usually issued in the amount of the contract. Its cost varies with the type of work, the size of contract, and the surety, but averages about 0.75% of the contract price.

Payment Bond. The Payment Bond guarantees that the contractor will pay all labor, subcontractors, and suppliers. Separate payment bonds are generally required only on public works projects. It generally will have a face value less than 50% of contract price if issued separately.

The cost of bonds is included in an estimate and will be charged to the client as part of the project cost if the contractor gets the contract. If he does not, he will not be reimbursed for the bid bond or any expenses incurred in obtaining surety assurance of willingness to provide performance and payment bonds. These expenses become part of general overhead.

Taxes. A contractor who accepts work in a number of states or countries will be subject to many different taxes and tax rates. On public works projects, the contractor will be exempt from certain taxes. For these reasons, most contractors prefer to separate taxes from the base costs against which they are levied so that experience data on costs are not polluted.

State Income Tax. More than 40 states have a state income tax and each has its own rate structure for business taxation. If a contractor's operations are confined to a single state, this cost item probably would be combined with federal income tax in the overall profit planning phase. A contractor with a multistate operation may choose to estimate state income taxes and include them as a job indirect cost.

State Business Tax. A number of states have a tax that is levied on every business. It is usually based upon business volume in that state.

Sales Taxes. Most states and many municipalities have a sales tax on material and equipment purchases.

Use Taxes. Occasionally, a state or municipality will levy a tax on each item of equipment brought into that area for use on a contract.

City Tax. Some communities have enacted legislation which taxes any business or construction project. Its purpose is to gain reimbursement from the business and its employees for use of city facilities or services since many employees may not live in the area of the project and not otherwise contribute to upkeep of city structures.

Advalorem Tax. Any tax based upon the value of an item such as a property tax on structures or equipment.

Foreign Taxes. On overseas work a contractor is subject to all host country laws, including those involving taxation, while still being subject to certain taxes by the United States or the business' home state.

COSTING OF JOB INDIRECTS

Job indirect items will have a total cost that is some combination of mobilization (transport and installation) cost, period costs, one-time costs, and demobilization (removal and transport) cost. For example, if a warehouse is to be constructed to support the project, its total cost is the sum of the costs of constructing it, periodic maintenance and operating costs, and the costs of dismantling it less any salvage value. For personnel the costs are periodic costs multiplied by the number of periods of employment. Support items for the personnel are otherwise accounted for. Bonds and permits are typical one-time cost items. Furniture and other equipment are best handled on a monthly rental basis since they can be reused on other projects. However, do not forget the costs of transporting these items to and from the project. Figure 14-2 is a sample worksheet for use by estimators in itemizing job indirect costs.

MARKUP

After all costs directly attributable to a job have been itemized and costed, a markup is applied to reflect general overhead, profit, and contingencies. Normally, the estimator will not apply the markup; it is a function of management and is accomplished after they have thoroughly reviewed the estimate with the estimators.

General Overhead

General overhead costs, also called main office expenses, are intended to include all those expenses related to the operation of the home office which cannot be directly tied to a given project so are distributed over all company projects. General overhead categories include

Management, clerical, and professional staff salaries and fringes
Estimating costs
Company sponsored training programs

Job Indirects Worksheet

Code	Item	One-time costs	Mobilization costs	Period costs				Demobilization costs	Total
				Period	Number of periods	Cost per period	Total		

Fig. 14-2 Job indirects worksheet.

Licenses, fees, professional literature
Company-wide insurance
Consulting fees (legal, CPA, etc.)
Centralized data processing service
Profit sharing programs and bonuses
Home office building rental, depreciation and maintenance
Utilities and communications
Home office vehicles and aircraft
Advertising and public relations
Office supplies, equipment, and printing
Interest payments not otherwise charged

As a general rule, the ratio of main office expense to volume of work decreases as the volume increases. The range is 2 to 20% of volume with a median value of about 7.2%.

To be realistic a contractor should project general overhead office expenses for the forthcoming year based upon the past year's experience adjusted for inflation and expected changes in contract volume. This total then should be related to the projected contract volume to establish the average charge rate to be used on the company projects in the coming year. The contractor may choose to charge more or less than this average rate on a given project if the home office contribution to that project is significantly more or less than average. In practice the rate would be a multiplier of either direct labor, all direct costs, or all project costs (direct plus indirect) depending on the contractor's desires for handling it.

Profit

In Chap. 3 we discussed return on investment and stated that a contractor is entitled to a return on his or her investment in the business. We went on further to say that a contractor with significant owner's equity with respect to contract volume should measure return against owner's equity as reflected on the balance sheet for the company. Owner's equity consists of contributed capital (or funds generated through sale of shares of stock in the case of corporations) plus retained earnings. A fair rate-of-return is a function of other rates-of-return available to the contractor (using these funds) and the risk incurred by being a contractor. The very lowest, or risk-free rate-of-return, is that available by purchase of US Government insured savings accounts. But, contracting is not risk-free so a contractor should get more than the risk-free rate—in fact, contracting is a very risky profession so he or she should get much more. We pointed out that the average rate-of-return after taxes for a number of major construction companies in the 1970s was about 15.6% and the before-tax rate-of-return was about 30%. Thus, lacking any better guidelines, it appears that a 30% before tax rate-of-return is a good base figure to use. In periods of high inflation and high prime interest rates (above 10%) higher rates-of-return are certainly justified.

Highly leveraged contractors (those with small owner's equity as compared to contract volume) will probably choose to handle profit strictly as a judgmental markup (such as 5 to 10% on subcontract items and perhaps 10% of his or her own resource costs).

Contingencies

A contractor is always concerned about the accuracy of an estimate and will organize the estimating function to insure that the estimates produced are as accurate as possible considering the known factors of the project. Unfortunately, on some projects, a contractor is asked to provide a firm-price bid despite the fact that project definition is somewhat vague or job conditions are likely to create problems. In these cases, the contractor must consider adding another factor to the markup: the contingency factor. As its name implies, the contingency factor is another markup or added amount to account for the unknowns in the project which make the project particularly risky. There is no solid guidance available for the size of such a markup. The contractor must decide what additional costs can be reasonably expected for items possibly overlooked or underevaluated and for which he or she has full responsibility. It must not be an unreasonable amount or the bid will be noncompetitive. Adding too little, on the other hand, may erode profits unreasonably.

Contingency amounts are not added to every bid since the data used in developing the estimate assumes average conditions, not ideal conditions, so there is a certain level of error built in. The amount of contingency added will tend to be inversely proportional to the amount of subcontracting used on the project and to the completeness of the plans and specifications. Chapter 16, Probabilistic Estimating, describes an approach for bid preparation which incorporates contingency considerations into the estimate of each work package.

Markup Example

The following example illustrates an approach for developing a usable markup figure.

example

A contractor projects a contracting volume of $1 million for the coming year and general overhead expenses of $70,000. The latest Balance Sheet shows that owner's equity is $100,000. The contractor wishes to determine a multiplier to use on total project costs (all direct costs plus job indirects).

Reasonable profit = (30%) × (Owner's equity) = $30,000 for year

General overhead   = $70,000 for year

Estimated project costs = (Volume) − (GOH + Profit) = $900,000

$$\text{Markup rate} = \frac{(\text{GOH} + \text{Profit})}{\text{Project costs}} = \frac{\$100,000}{\$900,000} = 11.1\%$$

Note: Any contingency markup would be added to this amount.

Cost Inflation Markup

Chapter 3 discussed the problem of inflation and its effect on construction costs. It is suggested that estimates be prepared using costs current at the time of estimate preparation and then adjusted within each work package for inflation once the project schedule is outlined. If a computerized, combination estimating and scheduling program is used, this can be handled automatically if escalation factors are incorporated into the Work Package Collection Sheets (see Fig. 15-4).

SUMMARY

Figure 14-3 is a diagram representing the components of a bid as discussed in this and earlier chapters. The bid is much like a jig-saw puzzle whose final picture is dependent upon the full inclusion and proper arrangement of the many parts. If properly prepared, a bid package can be readily reviewed by others and quickly adjusted if mistakes are found or planning decisions changed. Like a good set of accounting records, it provides a complete audit trail for its entries. Liberal use of collection sheets, such as the Work Package Collection Sheets used in Chap. 13, backed by worksheets containing supporting assumptions, calculations, and sketches, is extremely important. Collection Sheets are summarized by means of Summary Sheets, an example of which is shown in Fig. 14-4.

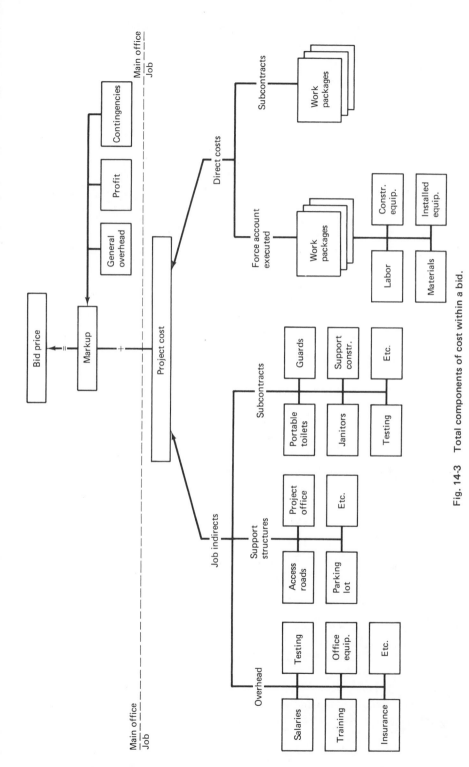

Fig. 14-3   Total components of cost within a bid.

271

```
┌─────────────────────────────────────────────────────────┐
│                    ESTIMATE SUMMARY                        │
├─────────────────────────────────────────────────────────┤
│  Project _____   Bid date _____   │
│                                                            │
│  Location _____   Estimator _____   │
│                                                            │
│  Client _____   Contract time _____  │
├─────────────────────────────────────────────────────────┤
│          Direct Cost Work Packages                         │
│               Permanent materials    $ _____            │
│               Materials and supplies    _____           │
│               Installed equipment       _____           │
│               Labor                      _____           │
│               Equipment operation        _____          │
│               Subcontracts               _____          │
│                  Total Direct             =======          │
│                                                            │
│          Job Indirects                                     │
│               One-time costs             _____           │
│               Mobilization costs         _____           │
│               Period costs               _____           │
│               Demobilization costs       _____           │
│               Subcontracts               _____           │
│                  Total Job Indirects     =======           │
│                                                            │
│          BID                                               │
│               Direct costs               _____           │
│               Job indirects              _____           │
│               General overhead           _____           │
│               Contingencies              _____           │
│               Escalation                 _____           │
│               Profit                      _____          │
│               BID AMOUNT               [_____]          │
└─────────────────────────────────────────────────────────┘
```

Fig. 14-4   Estimate summary worksheet.

REVIEW

Exercises

1. A contractor projects a contracting volume of $5 million and main office expenses of $350,000 for the coming year. Her Balance Sheet shows $80,000 of contributed capital and $600,000 of retained earnings. Calculate her markup on project costs.

2. Prepare a listing of job indirect items (without costs) for construction of a small apartment complex in your community. Remember that this is a relatively simple project so don't overdo it.

Questions in Review

1. Distinguish between general overhead and project overhead (job indirects).

2. What is the general range of general overhead expense as a percentage of contract volume?

3. A contractor has an open-end contract with a testing laboratory for services to all projects. Should the costs of tests performed on a given project be treated as a general overhead, job indirect, or other category of cost?

4. What is the difference between builder's risk and public liability insurance?

5. Describe types of taxes that a contract may be subject to.

6. Why should taxes be handled separately from the items taxed?

7. Who pays for bonds?

8. What determines a fair rate-of-return?

9. Should the same markup be used on all projects?

Discussion Questions

1. It is often said that a contractor should be able to estimate the cost of a project within 5% of actual cost if complete plans and specifications are available. If these plans and specifications are less complete, the potential margin of error increases. Do you think that profit markup should be related to these margins of error?

2. Some contractors have been known to intentionally submit a bid that includes no profit and which may not include complete coverage of general overhead costs. Why would a contractor choose to do this?

PUTTING IT
ALL TOGETHER

15

INTRODUCTION

The general process for development of a bid was described in Chap. 5. In Chaps. 6 through 14, the individual resources of construction and other components of cost were discussed. In this chapter these chapters will be reviewed through an outline description of the development of a bid for an industrial project.

THE PROJECT

For this example the small industrial project sketched in Fig. 15-1 will be used. The project is to be located on a currently undeveloped tract of land adjacent to a river, a highway, and a railroad. The project includes development of the site itself—clearing and grubbing, drainage structures, roads, ramps, parking, fencing, and a railroad spur. Two buildings are part of the project. The first is an administrative building; the second a structure housing process equipment used in manufacturing the products of this plant. In addition, there is a raw materials delivery system which includes a railroad car unloading hopper, conveyors, and an intermediate storage tower. For purposes of this problem it is not important that dimensional details, specifications, or types of products be known. Obviously, hundreds of detailed drawings would be required for the total project, but our involvement in such detail is not necessary to illustrate the general procedures for bid development.

DEVELOPMENT OF THE BID

Planning

As shown in Fig. 5-5, the first effort in bid development is full review of the contract documents, a visit to the construction site and

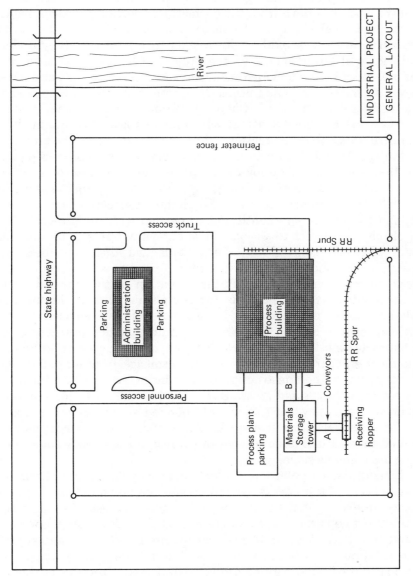

Fig. 15-1   Example industrial project.

275

area of the project, and development of the Work Plan. This is a task of the Project Team appointed by management to oversee bid development and later execution of the project should the bid be successful. Heading this team is a project manager who will coordinate the work of its members who are drawn from the estimating, scheduling, procurement, cost control, engineering, and construction groups. First, the entire team must review the bidding documents to ensure that they understand the total requirements of the project. The plans and specifications will describe the included facilities but careful attention also must be given to the General and Special Conditions since they include many items with cost implications (Chap. 4) and may include some, such as milestones, which directly affect scheduling, choice of crews, method of construction, or other matters. In addition, it is entirely probable that the clients will choose to order the major items of process equipment themselves so scheduling of the project must be tied to expected delivery dates of these items. Certainly, visits to the project site are in order to further understand project requirements, check the labor situation, and to identify any constraints associated with the project. The project may be subject to various jurisdictional controls as to environmental pollution constraints.   The clients and their engineers are responsible for incorporation of design features which accomplish desired environmental control during operation of the finished plant but the contractor is responsible for compliance with environmental regulations during construction. For example, the river adjacent to the plant may be a source of water for some municipalities downstream so the contractor must ensure that earthmoving or other operations do not contaminate the stream with sediments, oil spillage, or other pollutants. This may require the construction of a drainage system which incorporates sediment basins and grease traps. Additionally, the construction equipment yard may have to include bermed structures for containment of potential oil or fuel spills. All of these facilities cost money and take time so it is essential that the project team identify their need during the planning stage.

In addition to pollution control there may be other actions required because of the site. Access to the site may be a problem. Limited capacity bridges, potential traffic safety problems near the plant entrance during construction, remoteness of the site from the labor market and countless other problems may require investigation and appropriate solutions incorporated into the Work Plan. For example, to reduce auto traffic and parking problems during construction, as well as to save fuel and entice a distant labor market, the decision may be made to utilize busses to transport workers to and from the sources of labor.

Once items of responsibility in addition to those contained specifically within the contract documents have been identified, an overall plan for execution of the permanent facilities can be developed. To do this the project is divided into its major work packages. The next step is to make a determination of those packages to be accomplished with force-account resources and which to subcontract. For this project possible candidates for subcontracting are fencing, curbs and gutters, paving, landscaping, and interior finishes. If the clients have chosen to be responsible for procurement of certain components, this is so noted. In addition, contractor-procured items which have potentially long procurement lead times will be listed. This list will go to the procurement division for estimates of both cost and delivery time. Finally, a logic diagram (Fig. 15-2) is drawn for the project which interrelates the major work packages and their sequence of execution. This logic diagram is just that—it is not a schedule because, at this point, durations are not known. Also, it is very broad in its work package definition; it is the framework for a later schedule. With further breakdown of the work packages during estimating and scheduling into their component packages plus addition of activities such as prefabrication, ordering lead time, and delivery time, the Level II network is expanded to the greatest level of detail, the crew task level, which for this example is assumed to be Level V. Figure 15-3 is an example of a portion of such an expansion. On all logic diagrams it is appropriate to make annotations which indicate any milestone dates, identification of subcontracts, or other pertinent items of information that may be needed later by estimators or others.

Thus, at this point, the project team has developed the Work Plan for the project. It is the outline or skeleton of the project which will be fleshed out by estimating, procurement, and scheduling personnel. In summary the Work Plan consists of

1. Identification of component work packages
2. A logic diagram of site level major activities
3. Identification of force-account components
4. Identification of major subcontracts and procurements
5. Other planning decisions or goals

Procurement Actions

With the decision having been made to subcontract certain work packages and assuming that certain major procurements must be handled by the contractor, the Procurement Department goes into action.

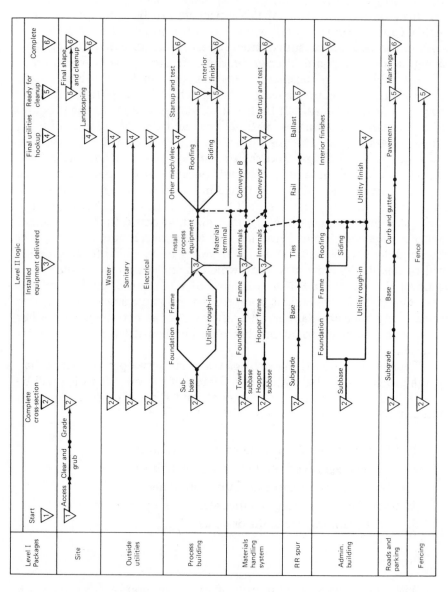

Fig. 15-2 Level II logic.

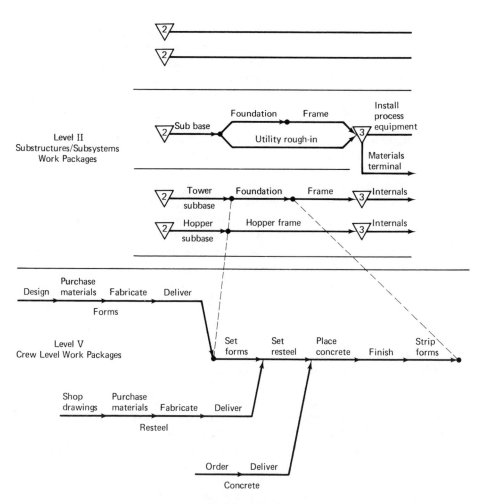

Fig. 15-3   Expansion of Work Packages from Level II to Level V.

For subcontract items they will attempt to obtain bids prior to the general contract bidding deadline. To do this, the Project Team must furnish them extracted drawings, specifications, and other general contract documentation needed to develop the subcontract bid packages. Similarly, documentation is developed for vendor-supplied major items on which quotations are needed. If time is not available to solicit firm bids and quotations prior to the general contract bidding deadline, the Procurement Department, in conjunction with the Estimating Department makes an estimate of costs, lead times, and durations. If the major process equipment is client-procured, no action on these is required by Procurement Department unless the client will require specific services, such as expediting or factory inspection. In such case, the Procurement

Department must provide an estimate to the Project Team of the cost of these services to incorporate in the bid. For all other work packages, the primary responsibility for determining cost, resource requirements, and durations belongs to the Estimating Department.

Estimating Actions

The Estimating Department has several responsibilities included within their part of bid preparation, namely

1. For all force account work packages, to determine resource requirements (permanent materials, materials and supplies, installed equipment, labor, and construction equipment), expected duration of the work package and its cost
2. For subcontract packages, to develop an approximate estimate of cost to be used in evaluation of bids (optional)
3. To list, quantify, and cost all job indirects
4. To consolidate and summarize all project-level costs to include subcontract and vendor-fabricated items

Work Package Estimating. The work packages developed as part of the Work Plan provide the framework for estimating. For this project, the Level I packages included major systems and structures such as administration building, process building, materials system, and site work. While an estimator could produce an approximate estimate of cost for each system or structure based on factors such as cost per square foot of similar structures or cost per unit of output of similar plants, such an estimate is subject to considerable error so the estimator, time permitting, will estimate at the lowest level of work package definition possible. Ideally, this is the crew level and would include actions such as those identified in Fig. 15-3. Working at this level of detail, the estimator is more likely to identify all cost elements. Assuming that the estimator is preparing an estimate for each crew task, he or she will utilize Collection Sheets of a design similar to those used in Chap. 13. Another of these forms is shown in Fig. 15-4. This one differs from the others only in that entries are set up for computer input. Note that extensions have not been made. This will be done by the computer based upon its handling of the low, target, and high productivity rates assigned. This subject will be discussed more in the next chapter. Also note the use of escalation rates. The computer, in processing the estimate, can also schedule the project and incorporate escalation.

Job Indirects. The many possible job indirect items were listed and discussed in Chap. 14. As suggested in that chapter a detailed

WORK PACKAGE COLLECTION SHEET

| System/Structure Identifier | 5 1 • 2 2 0 | Crew Level Work Package Identifier | 0 3 • 1 4 3 | Description | CONCRETE PLACEMENT (PUMPED) HOPPER TOWER FOUNDATION |

Productivity

Base Unit for Productivity CY CONCRETE

Total Quantity Base Unit 100

Permanent Materials (PM)

Resource Code	Description	Unit	QTY	Unit Cost Low	Target	High	Extension
1 3 3 5	CONCRETE, 3500 PSI	CY	103*		36 -		
	*ALLOW 3 CY WASTE						

Other Materials and Supplies (M&S)

Installed Equipment (IE)

Crew Labor (L)		NR.	Cost/Hour Low	Target	High	USE Hours
0 2 9 1	FOREMAN	1		10 90		
0 2 9 3	LABORER	4		10 40		
0 2 9 2	FINISHER	1		12 85		
0 9 1 3	PUMP OPERATOR	1		13 25		

Equipment Not Charged As Job Indirects		NR.	AVAILABILITY Periods	$/Period	Hours	USE		$/Hour
7 3 1 1	VIBRATORS, GED	2 @	NOT APPLICABLE	×	( ) =	×	3.25	=
7 3 1 2	FINISHER, GED	1 @	NOT APPLICABLE	×	( ) =	×	4.50	=
8 3 3 1	CONC. PUMP, TRK MTD	1 @	NOT APPLICABLE	×	( ) +	×	75.00	=
		@		×				=

Duration (Crew-Hours)

	Low	Target	High
	7	8	10

Escalation Rates (%)

Materials 1 1
Labor 0 8
Equipment 1 1

Notes

Cost Summary: PM =
M&S =
IE =
L =
CE =
TOTAL =

Fig. 15-4 Work Package Collection Sheet for concrete placement.

281

checklist should be used to best ensure that some element of job indirect cost is not ignored. Many of these cost items are one-time expenditures, such as bonds and mobilization costs. Others are a function of duration of the project, such as wages of the construction manager, drinking water, and office rental. Since the length of the project is not known while the estimate is being prepared, some of the job indirect items cannot be accurately estimated. However, if the job indirect listing is prepared in a fashion to isolate those costs which are a function of project length, their costs can later be determined when the trial schedule is developed. Again, if a computerized method of estimate preparation is being used, the computer can take care of these calculations automatically.

      Completing the Project-Level Estimate.   Total project-level costs are a combination of direct work packages (both force-account and subcontract) and job indirects. Thus, it is appropriate to provide an overall summary of these costs through a summary printout or summary sheet prepared manually. This summary should summarize the costs by resource category and, if a probabilistic estimating approach is being used (Chap. 16), give the frequency distribution curve of project-level costs and project length for use by management during markup.

      Markup.   The project team is responsible for coordination and final review of the estimate and trial schedule prior to its presentation to management. Once management is satisfied with the project-level estimate they will make decisions on final bid price to incorporate profit, a distributed share of general overhead (main office) expenses, and consideration of contingency.

SUMMARY

The bid for a construction contract is a serious commitment for a contractor should he or she win the contract. Obviously, great care should be exercised in its preparation to ensure consideration of all included costs. This is best done through an approach utilizing work packaging. Another requirement is a complete chart of accounts which can be used for all project control actions. This chart of accounts is the language of the system and is the key to computerization. Of course, a comprehensive data base is also essential for success. Utilizing the approach suggested in this text, an estimator best ensures development of a complete and accurate bid in a format which directly supports scheduling, budgeting, and cost control. This format also can be readily adjusted should addenda to bidding documents be received or should later changes be made in the contract.

REVIEW

Exercises

1. Expand a Level II work package for installation of an underground water line into its crew level packages.
2. Assume you are preparing a Work Package Collection Sheet for placement of reinforcing steel in a large floor slab. List by category the resources you should account for on the collection sheet. Do not include quantities or costs. How would the list change if the work was in an elevated location?

Questions in Review

1. In what way does work packaging for estimating purposes facilitate scheduling, budgeting, and cost control during construction?
2. Why should a trial schedule be prepared during the bidding phase?

Discussion Questions

1. Considering the approach recommended by this text for formulation of a bid, how do you recommend that individuals be developed as effective cost engineers within a large contracting organization?
2. On the staff of a large construction project, it is quite common for a Project Engineering Group to be formed which is responsible for on-site design tasks (such as form design), surveying, review of design drawings prepared by others, surveillance and inspection of work in progress, resolution of engineering problems, and construction document control. On many staffs, scheduling and cost control is also part of this group; on others, all cost engineers are in a group parallel to the Project Engineering Group and this group reports directly to the construction manager. If you were a construction manager, which arrangement would you prefer?

CONSIDERATION OF VARIABILITY THROUGH PROBABILITY

16

INTRODUCTION

There are many variables in any construction project. Some variables, such as labor productivity, have a rather broad range of possible values. Others, such as wage scales, tend to have narrower ranges. For each variable, it is the task of the estimator to select a target figure applicable to the project being bid. The probability of selecting the correct value from within the range is almost zero so what the estimator is really trying to do is select a value which has a high probability of varying only slightly from the real value. Overall, the bid becomes a composite educated guess composed of many component educated guesses on the part of the estimators. What then is the composite effect of all these guesses? Add to this question some others. Did the estimators properly evaluate site conditions? Did the estimators use shortcut methods which may be unrealistic in this situation? What did the estimators leave out? The final question is: How much money will the contractor risk losing if he or she were to submit a bid based on the raw estimate of cost? The point is that the bidding process carries with it an inherent risk of being wrong, usually on the low side, and the contractor must modify the bid to adjust for this risk. This is usually done by adding contingency.

How much contingency to add is a major decision. If the plans and specifications appear complete and understandable or if a major portion of the work is subcontracted to reliable subcontractors whose bids have been received, the contingency addition need not be more than 1 to 2%. However, if the plans are conceptual or poorly done, if there are features of the project that are unsettled at time of bidding, or if short-cut procedures were necessary in preparing the bid, a contingency markup of 10%, 20%, or higher may be in order. Obviously, if the markup is too low the contractor may be risking profit, general overhead, or even the company's retained earnings. If the markup is too high the bid may be noncompetitive. Unfortunately, regardless of the markup applied, the contractor has really done nothing more than add contingency based upon some rule-of-thumb he or she considers applicable. In this chapter, an alternate approach to handling risk will be described. The approach utilizes probability.

APPLICATION OF PROBABILITY TO VARIABILITY

A cost estimate is composed of numerous line items to which an estimator eventually applies a cost. A typical line item may be a composite of labor, equipment, and material. Uncertainties include labor wage scales, labor burden costs, labor productivity, material quantities and costs, and equipment costs. Yet, for each work item, the traditional approach is to arrive at a single target cost. The probability of the actual costs being exactly that amount are essentially zero. However, the probability that they will be within a range around this value is significant. Assume, for example, that a work item is estimated to have a most probable cost of $100,000. If the estimator were to assume the worst combination of the variables, he or she may find that it would cost $150,000. Under the best of conditions, he or she may determine that it would cost only $80,000. In effect, the estimator has determined that there is close to 100% assurance that the cost will be between $80,000 and $150,000. Having chosen $100,000 as the most probable cost, the estimator has conveyed his or her feeling that the greatest concentration of probable cost is near $100,000 with lesser probabilities toward the extremes. In doing this, the estimator has established a probability profile for this line item which is a composite profile of the probability profiles of the individual variables in the line item. At this point the estimator is still combining evaluation of a number of variables at once but not to the extent that they are combined when overall contingency factors are applied to a total project. Now, if the estimator were to back up and consider each of the variables in each line item separately and statistically determine their composite effect on the total line item cost, he or she will closely approach optimum consideration of variability and risk.

The use of probability in variability analysis is nothing new. Typical finance texts describe methods of applying probabilistic methods to analysis of investment portfolios, credit policies, and other financial actions. Its use in construction has been limited, although is growing slowly in popularity.

To explain probabilistic estimating a simple example will be used. In this example a project consists of only two components, Package A and Package B, so that the total project cost is the sum of the cost of those two packages. As the estimators examine Package A, they develop a cost profile for it. For example, as shown in Fig. 16-1(a), the estimators feel that Package A will cost between $6 and $9 million dollars but that the most probable cost is $8 million dollars. Looking at it another way they conclude that if the package were constructed several thousand times, the outcome in cost would show a distribution that ranges between $6 and $9 million with the most frequently occuring cost being $8 million. The shape of the distribution curve is shown as triangular in this example for ease of explanation but other shapes could be assumed as more realistic for the variable under construction. For Package B, the variation is shown in Fig. 16-1(b). Here the estimators have concluded that the possible range of cost is between $3 million and $5 million dollars with $4 million most probable. It follows that the total cost of the project can be assumed to range between $9 million ($6 million low for Package A plus $3 million low for Package B) to $14 million ($9 million high for Package A plus $5 million high for Package B).

The next step is to develop an overall probability curve for the total project. To do this, first recognize that the area under each of the two curves represents 100% probability of occurence and that we can calculate the probability of a cost being between any two values of the curve by relating the area under the curve between those two points to the total area. Referring again to Fig. 16-1(a), the area under the curve between $6 million and $7 (Area A1) is 12.5 in terms of the small squares of the graph. Similarly, the area between $7 million and $8 million (A2) is 37.5 squares and between $8 million and $9 million (A3) is 25 squares. The total area under the curve is 75 squares. Thus, the probability of the cost being between $6 million and $7 million is numerically equal to

$$\frac{12.5 \text{ squares}}{75 \text{ squares}} = 0.17 \text{ approximately}$$

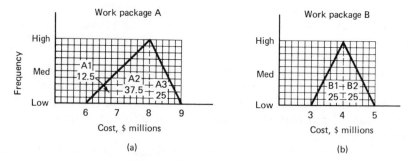

Fig. 16-1   Frequency distributions for Work Packages.

By similar calculations the probability of the cost being between $7 million and $8 million would be 0.50 and between $8 million and $9 million is 0.33. For Package B, the probability of the cost being between $3 million and $4 million (B1) is 0.50 as is its probability of being between $4 million and $5 million (B2). Next, let the midpoint cost in each $1 million range represent that range. While this is only an approximation, it would be more accurate if the ranges were reduced in width. For example, the range between $6 million and $7 million in Package A is represented by $6.5 million. The resultant probabilities are

Package A	Increment (million $)	Area	Probability	Cost (million $)
(A1)	$6–$7	12.5	0.17	$6.5
(A2)	$7–$8	37.5	0.50	7.5
(A3)	$8–$9	25.0	0.33	8.5
			1.00	
Package B				
(B1)	$3–$4	25.0	0.50	3.5
(B2)	$4–$5	25.0	0.50	4.5
			1.00	

The next step is to identify all possible combinations of cost using the selected ranges. These are

Possible Combinations	Total Cost	Combined Probability*
A1 + B1	$10.0	0.08*
A1 + B2	$11.0	0.08
A2 + B1	$11.0	0.25
A2 + B2	$12.0	0.25
A3 + B1	$12.0	0.17
A3 + B2	$13.0	0.17
		1.00

*Found by multiplying probabilities of the two increments in each combination and rounding off.

Noting that more than one combination yields a given total cost in some cases, the information can be consolidated to give the total probability of each project cost. In addition, the probability of each cost not being exceeded can be calculated by cumulatively adding these probabilities. This is summarized below

Project Cost	Total Probability	Probability of not Being Exceeded
$10.0	0.08	0.08
$11.0	0.33	0.41
$12.0	0.42	0.83
$13.0	0.17	1.00

From this data a curve is drawn which relates probability of a cost not being exceeded to total project cost. This is shown in Fig. 16-2. This curve gives the estimator a basis for selecting a bid price. For example, as shown on the graph, a total project cost which has a 50% chance of not being exceeded is $11.2 million. Similarly, the total cost which has a 75% chance of not being exceeded is approximately $11.75 million. Thus, by using a probabilistic approach to bid development the estimator can associate a definite risk with each bid price and on this basis select the price to bid. For example, company policy might be to never bid a price which has greater than a 40% chance of being overrun. If this policy is adhered to with all projects, the company has reasonable assurance of generating overall profit even though some projects will overrun.

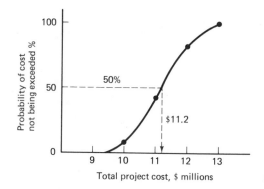

Fig. 16-2 Project cumulative probability curve.

Refinements to the Process

In the example given a number of simplifying assumptions were made to explain the method. One of these was that the distribution is triangular. While the exact shape of the distribution can never be precisely predicted, it may not be triangular in the opinion of the estimators. However, other distribution shapes can be handled by breaking the assumed distribution into equivalent rectangular areas as shown in Fig. 16-3 and handling as before.

Another approach is to use a modified triangular distribution and

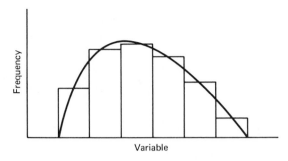

Fig. 16-3 Converting a frequency distribution to equivalent rectangular areas.

to add one more assumption by the estimator, that being the probability that the target estimate will not be be overrun. Figure 16-4 is such a distribution which results from an estimator's decision that the work package will cost somewhere between $130,000 and $170,000; the most probable cost is $160,000 and the probability of $160,000 not being exceeded is 67%. The ordinates of the two component triangles are a function of the probability assigned so that the area of the left triangle is 67% of the total area while the right triangle is 33%. Use of this form of distribution is particularly suited for computerized analysis since equations for each of the lines bounding the distribution are easily written in terms of the assigned quantities: low value (L), high value (H), target value (T), and probability of underrunning target (P) enabling the equations of the cumulative probability curve to be written as shown on the figure.

Computerizing the Process

Handling a project with many unknowns by the longhand process is completely impractical but many unknowns are readily handled by computers. One method uses the double triangle distribution approach just described. For each variable the estimator assigns a low, high, and target value plus the confidence factor in terms of probability that a variable will not be exceeded. This automatically describes the shape of cumulative probability function for each as shown precisely in Fig. 16-4. Next, the computer makes repeated costing runs of the project. In each run, as it encounters a variable, it selects a value for it. In effect, it randomly selects a value between zero and 100 to represent the ordinate of the cumulative probability function and from that value calculates the variable. This technique automatically accounts for the probability distribution assumed by the estimator. The total project cost for that run then becomes the cost of fixed items plus costs of variable items selected during that run. After many runs are made, an overall cost profile for

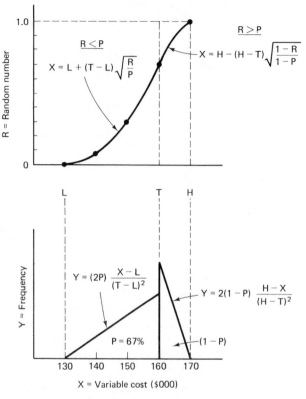

Fig. 16-4   Double triangle distribution.

The equations shown in the figure are:

For $R < P$:
$$X = L + (T - L)\sqrt{\frac{R}{P}}$$

For $R > P$:
$$X = H - (H - T)\sqrt{\frac{1 - R}{1 - P}}$$

$$Y = (2P)\frac{X - L}{(T - L)^2}$$

$$Y = 2(1 - P)\frac{H - X}{(H - T)^2}$$

$P = 67\%$

$(1 - P)$

X = Variable cost ($000)

the project and a corresponding cumulative probability function can be drawn.

Developers and users of a computerized program of the type outlined must be very careful in the writing of programing equations to not introduce unrealistic conditions. For example, if the cost of concrete at a point in time is one of the variables and it appears in a number of the equations, it should not be ranged separately for each occurrence since the computer would assign a different price for the same item each time. On the other hand, worker productivity on one work package may be completely independent of productivity on similar packages. Also, some variables will tend to vary together, such as inflation rates. While materials, labor, and equipment costs may not inflate or deflate at the same rate, it is unlikely that one will inflate at the highest rate within its range while another inflates at a low rate in its range.

Contingency cost considerations other than those directly associated with conventional work package and indirect cost items can be included in the computerized program just described. Two representative equations are these:

$$\text{Strike cost} = (\text{Probability of strike}) \times (\text{Strike cost/day})$$
$$\times (\text{Length of strike})$$
$$\text{Dewatering Cost} = (\text{Probability of groundwater})$$
$$\times (\text{Cost of dewatering operation})$$

In each case the estimator can range any or all of the variable items. As a practical matter, even though there are hundreds or thousands of variables in a project, the estimator should range only those variables which have a potentially significant impact on project cost and use conventionally selected target figures for all others. McDonnell Douglas Automation Company (MCAUTO) has a commercially available "Range Estimating" program and they recommend that elements selected for ranging be limited to those whose variation could affect project cost by more than 0.1%. In addition, they believe that a typical project will seldom have more than 50 items in this category.

SUMMARY

The use of a probabilistic approach to cost estimation has many advantages over the traditional single value type of estimate.

1. Probabilistic inputs encourage detailed consideration of each variable separately. Their use tends to bring all members of the project team into the estimating process.
2. The variables are properly weighted in the overall analysis.
3. The display outputs have meaning to management and make more meaningful consideration of contingency possible.
4. As project definition proceeds from conceptual to definitive levels, it is possible to continually refine variables and develop new cost profiles.
5. Sensitivity analysis to show effect of specific project changes can be readily handled.
6. On cost-plus proposals a contractor can provide the client with a more realistic picture of potential project cost range.

REVIEW

Exercises

1. A project is composed of two work packages. Package A is expected to range in cost from $3 million to $6 million with $4 million the target cost. Package B

ranges from $2 million to $6 million with $5 million the target cost. What bid price gives a 70% assurance of not experiencing an overrun?

2. Draw a double triangular frequency distribution curve for a work package if the low expected value is $300,000, the high value is $700,000, and the probability of cost not exceeding $400,000 is 40%.

Questions in Review

1. What is the area under a frequency distribution curve?

2. Why is it correct to say that the probability of a project cost being a specific amount is zero? In what way should probability of cost be expressed?

Discussion Questions

1. There is a tendency among engineers and business managers to somehow attach numbers to everything involved in decision making so that final options can be presented in numerical form for comparison. The approach of this chapter attempts to provide a numerical approach to decisions which previously were ones of judgment. Which approach do you feel most comfortable with?

2. Many of the senior managers in construction have not had formal training or experience in the application of advanced techniques, such as range estimating, and are reluctant to try them. They point out that existing procedures have worked successfully for many years and all personnel are familiar with them. So, why change? How would you go about selling an old-line manager on a new technique such as range estimating?

ESTIMATION AND CONTROL
OF ENGINEERING COSTS

17

INTRODUCTION

For the design-build contractor, the estimation of engineering costs is as important as the estimation of construction costs. As with construction, the engineering contract may be fixed-price, cost-reimbursable, or some other form. Whatever the contract form, the design-build contractor will find it necessary to estimate design costs either for the purpose of advising the client for budgetary purposes or for inclusion in a fixed-price design-build contract. Thereafter the contractor controls these costs through schedules and other control techniques. It is the purpose of this chapter to describe a system for estimation and control of these engineering costs.

THE SYSTEM

General Approach

As with construction, a system for estimation and control of engineering costs must contain procedures for identification of component work packages, estimation of quantities of resource expenditures within each work package, scheduling of effort based on resource requirements and resource availability, and control of total effort through comparison of actual resource expenditure to budgeted. The primary resource involved in engineering is the employee-hour.

Work Packaging of Engineering

Work packaging of engineering follows the same breakdown as for construction except at the crew task level (see Chap. 5). With engineering, the crew tasks are no longer such things as setting of resteel and pouring of concrete, they are

Preparation of studies	Subcontract preparation
Development of flow diagrams	Preparation of reports
Architectural design	Requesting of permits
Structural design	Specification writing
Mechanical design	Expediting
Electrical design	Estimating
Instrumentation design	Scheduling
Other specialized design	Cost controlling
Drafting	Procedure writing
Purchase order preparation	Model construction

However, the objective is the same when estimating engineering tasks— to isolate for each work task the number of units of work to be performed, the resources required to perform that work, and the estimate of cost.

Resource Requirements

Unfortunately, the resources required for a particular engineering task, such as preparation of a drawing, vary over a much broader range than the effort required for a typical construction task, such as the placement of reinforcing in a foundation. Thus, an averaging approach is used in the handling of employee-hours for each work task category. For example, the range of employee-hours required for production of a structural design drawing may range from as few as 15 to as many as 200. Through past experience on structural work, the company may have learned that the average number of employee-hours per structural drawing is 120. This average figure is then used for all employee-hour estimates on structural drawing work. To illustrate, assume that the engineering effort is being estimated for the structural frame of a building. One of the work tasks included is preparation of design drawings. First, the estimator estimates the number of drawings that will be required for that frame. He or she then converts this to a total employee-hour requirement by multiplying this number by the average effort of 120 employee-hours per drawing.

While there are resource requirements other than employee-hours included in engineering tasks, such as office supplies, computer time, and other items, only the employee-hour requirements are treated as a resource. All other items are accounted for by a markup to the cost per employee-hour.

Development of the Cost per Employee-Hour

For each work task of the types listed earlier, a typical grouping of personnel is assumed to make up the basic staff directly involved in the action involved. This group includes the first-line supervisor plus the professional staff, technicians, and other subordinates. Not included in this group are higher level managers, word processing workers, and others whose efforts support more than one of the basic production groups. For this basic production group the total cost of salaries and wages is calculated. Employer contributions to social security, worker's compensation, and unemployment insurance also may be included in this calculation. Next, an estimate is made of the cost of travel, office supplies, reproduction costs, communications, equipment rental, and other costs attributable directly to the basic production group and this is added to the salaries/wages total. Then, an estimate must be made of all costs attributable to the engineering staff above the basic production unit level. These costs would include senior engineering staff, their secretarial staff, word processing centers, libraries, shared equipment, and similar cost items. The total cost of these items is distributed proportionally to the production units on the basis of population. Finally, a distributed cost to represent a proportional share of general overhead expenses must be added. These expenses include company level staff, support staff sections such as legal and personnel, building depreciation, utilities, company-paid fringe benefits not otherwise accounted for, etc. All of these costs are then added together and divided by the number of employee-hours in the production unit. In equation form

$$\text{Cost per employee-hour} = \frac{Q1 + Q2 + Q3 + Q4}{(N)(H)}$$

where: $Q1$ = Basic salary/wage cost per year of basic production group plus fringes

$Q2$ = Cost of supplies and other items directly attributable to the basic production group (annual rate)

$Q3$ = Distributed portion of other engineering staff costs (annual rate)

$Q4$ = Distributed portion of main office expenses (annual rate)

N = Number of personnel in basic production unit

H = Number of working hours per year

Milestones Within Work Tasks

At this point the procedure for determining an average employee-hour requirement per task and an average cost per employee-hour have been described. When combined with the estimated number of tasks to be performed, the estimator will have the basic input needed for the engineering estimate. However, it must be recalled that the estimate is but a first step in project control so some mechanism must be established to later permit comparison of actual performance to target. To provide this mechanism, milestones must be established within each work task that provide a measure of relative progress. For example, in the case of drafting, completion of a drawing may be measured as follows:

Title block and outline	10%
General layout	30%
Complete and ready for review	80%
Issue for construction	100%

All other engineering tasks are comparably structured using applicable milestones for the task involved and a percentage completion figure for each milestone. Having done this, a system is available for later control of engineering costs as is illustrated in the following example.

example:

A contract for engineering of a major facility is being estimated. At this time, the detailed estimate for engineering, drafting, and checking of drawings is being prepared. After review of the project requirements and review of historical data on drawing and employee-hour requirements for comparable projects, the following summary of estimated requirements has been generated:

Category	Number of Drawings	Employee-Hours/ Drawing	Total Employee-Hours
Flow Plans	10	80	800
Layouts	20	80	1,600
Civil/Structural	150	120	18,000
Piping	180	150	27,000
Mechanical	80	120	9,600
Electrical	50	90	4,500
Instrumentation	25	200	5,000
Contingency Reserve for Revisions			6,650
		Total	73,150

In a similar fashion, the employee-hour requirements for all other functions included in the project would be estimated in terms of employee-hours. For purposes of this example, assume that the totals are as follows:

Function	Total Employee-Hours	Cost/ Employee-Hour	Extended
Engineering/drafting	73,150	$23.40	$1,711,710
Procurement	3,200	21.00	67,200
Scheduling	800	21.00	16,800
Specification writing	400	21.00	8,400
Quality assurance	600	21.00	12,600
Regulatory compliance	1,200	21.00	25,200
Reports and studies	500	21.00	10,500
			$1,852,410

Engineering work is scheduled in conventional fashion and each activity is given its share of estimated employee-hours as a basis for control during actual execution.

Progress is reported in terms of milestones achieved within each activity and actual employee-hours expended. For example, assume that the milestones for drafting are as previously described and that drafting employee-hours make up 30% of the total employee-hours in the engineering/drafting function. Calculate the status of drafting given the following data

Total estimated (target) employee-hours, civil/structural	18,000
Portion allocated to drafting = (18,000) (30%)	= 5,400 TEH
Actual drafting employee-hours expended to date	= 4,500 AEH
Scheduled employee-hours to date	= 4,080 SEH

Status of drawings:

Stage of Completion	Number of Drawings at This Stage
Title block and outline	6
General layout	8
Complete and ready for review	10
Issued for construction	110

The earned employee-hours (EEH) (see Chap. 5) are calculated by multiplying the number of drawings at each stage by its milestone completion factor and the allocated drafting employee-hours per drawing (120 EH × 30% = 36 EH) as follows:

Title block and outline	(6 drawings) (10%) (36 EH)	= 21.6 EEH
General layout	(8 drawings) (30%) (36 EH)	= 86.4 EEH
Complete and ready for review	(10 drawings) (80%) (36 EH)	= 288.0 EEH
Issued for construction	(110 drawings) (100%) (36 EH)	= 3960.0 EEH
	Total Earned Employee-Hours	= 4356.0 EEH

This information is then utilized to complete the analysis of the situation at the end of the reporting period as follows:

$$\text{Overall \% complete} = \frac{\text{Earned employee-hours}}{\text{Target employee-hours}} = \frac{4356 \text{ EEH}}{5400 \text{ TEH}} = 81\%$$

Since earned employee-hours to date (4356) are less than actual employee-hours to date (4500) civil/structural drafting costs are apparently overrunning budget.

Since earned employee-hours to date (4356) are more than scheduled employee-hours to date (4080), civil/structural drafting is apparently ahead of schedule.

Order of Magnitude Estimates of Engineering Costs

With all types of estimates, there is a need for procedures to prepare quick approximations of engineering costs. Such estimates are needed during early planning stages and for running rough checks against more detailed estimates. A popular method for doing this is to develop plots of engineering costs as a percentage of total project cost for each type of project. In general, there is an economy of scale associated with engineering costs so that the percentage decreases with project size. A complex process plant designed from scratch may experience design costs approaching 30% of total installed costs, particularly

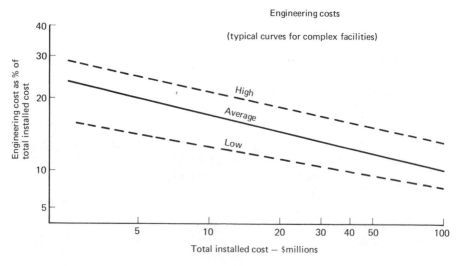

Fig. 17-1

if it is a pilot plant or one of relatively small volume, while larger plants of the same type may have design costs that are less than 10% (see Fig. 17-1). Facilities that can be designed around standard components, such as a fossil-fuel power plant or a water treatment facility, should experience design costs more in the range of 3 to 10%, depending on size. Building design costs tend to be in the same range.

SUMMARY

Engineering costs on a project can range from a few percent to more than 20% of total project cost depending upon the type and complexity of the project. Thus, engineering cannot be treated as a minor cost item. Procedures must be established which permit an engineering firm to make reasonable estimates of these costs during the proposal stage and to later control these costs as a key component of total project cost. Employee-hours are the logical control variable. Each engineering activity should carry with it an average employee-hour requirement based upon the company's historical experience. In addition, for each engineering activity, there must be internal milestones established which permit determinations of percentage completion of each activity.

REVIEW

Exercises

1. Using data in the chapter analyze the status (as to budget and schedule) of drafting of electrical design drawings given the following:

Title block and outline complete	4 drawings
General layout complete	3 drawings
Complete and ready for review	6 drawings
Issued for construction	22 drawings
Actual drafting employee-hours expended to date	= 750
Scheduled employee-hours to date	= 800

2. Develop a list of potential milestones within each of the following activities:
   a. Writing a procedure
   b. Engineering of piping

Questions in Review

1. Why are employee-hours the best basic control unit for engineering?
2. Why is percent complete calculated by dividing earned employee-hours by target employee-hours?

Discussion Questions

1. In this chapter's discussion of milestones for drafting, it was stated that 100% of the target employee-hours were earned for a drawing when it was issued for construction. This approach presumes that any later changes will be handled through the contingency employee-hour allowance included in the original estimate. Another approach is to allow about 90% for this milestone, leaving 10% to cover changes. Which approach is best?

2. The percentage complete of any work package is found by comparing physical progress with the latest estimate of work required. Is this more difficult to do with engineering tasks than for construction tasks?

ESTIMATING FOR
UNIT-PRICE CONTRACTS

18

INTRODUCTION

Unit-price contracts were introduced in Chap. 4 as a special form of lump-sum contract that is commonly used on earthmoving, highway, and other contracts where exact quantities of major work items cannot be defined by the plans and specifications. This chapter will illustrate the process of developing bids for such contracts.

THE BID FORM

As explained in Chap. 4, the bid package for a unit-price contract will contain a Bid Form that lists the work units that make up the total contract package. The choice of these work units is up to the client, not the contractor. It is the contractor's responsibility to account for all costs within those work items that are on the Bid Form—none can be added. Assume, for example, that a highway project to be contracted consists of clearing the right-of-way, basic shaping of the subgrade by cutting and filling along the right-of-way, placement of select base course materials, culvert installation, paving, and final dress and landscaping of the shoulders and ditches. The bid form for this project is shown in Fig. 18-1. In this case the entire project is to be accounted for within the seven line items listed. There are no entries for items such as job indirects or markup. Also, two items have been described as lump sum (LS): structures and landscaping. What these involve can be determined only by thorough review of the plans and specifications. The quantities shown on the bid form are estimated quantities by the design engineer. Actual quantities may be more or less but the quantities shown are the ones to be used in extending bid prices as will be shown later.

Item	Unit	Engineer's estimate	Bid price	
			Unit price	Extended cost
Clearing and grubbing	Acre	50		
Rock excavation	CY	50,000		
Common excavation	CY	200,000		
Select fill	CY	300,000		
Paving	SY	100,000		
Structures	LS	1		
Landscaping	LS	1		
			TOTAL BID	

Fig. 18-1   Bid form for a unit-price contract.

ORGANIZING THE ESTIMATE

As with all projects, the planners and estimators must thoroughly study the plans and specifications to identify all work packages and to develop the Work Plan. Estimating will still be accomplished by work packages but these individual work packages must be combined into groups corresponding to the seven bid form work items. For the project, the planners and estimators choose to catalog work packages within work items as follows:

Clearing and Grubbing
  Clearing and grubbing
  Debris disposal
  Detours
Rock Excavation
  Drilling
  Blasting or ripping
  Disposal of rock
Common Excavation
  Cutting and hauling
  Spreading and compaction
  Ditch shaping
Select Fill
  Purchase of material (royalty), load, and haul
  Spread, compact, and shape
Paving
  Forming
  Resteel placement
  Concrete production and delivery

Placement
Finishing
Curing
Shoulder, backfill, trim, and final compaction
Structures
Culvert installation
Interceptor ditches
Drop chutes for drainage
Landscaping
Final trim and cleanup
Seeding of slopes

With the work packages identified, preparation of collection sheets for each of them is accomplished. Also, job indirects must be itemized and costed and a required markup for profit and general overhead determined. For this project, assume that direct costs, job indirect costs, and markup are as follows:

Direct Costs	
Clearing and grubbing	$ 75,000
Rock excavation	622,000
Common excavation	210,000
Select fill	571,000
Paving	1,500,000
Structures	35,000
Landscaping	40,000
Total Direct Costs	$3,053,000
Job Indirects	300,000
Markup	370,000
Total	$3,723,000

Since job indirects and markup do not appear as line items on the Bid Form, these must be distributed over the seven items that do appear. The method of distribution is up to the contractor. As will be explained later in this chapter, he or she can choose to *load* some items. At this point in the explanation, let us assume that the choice is to distribute these costs proportionally. To illustrate, the amount to be distributed to clearing and grubbing is found as follows:

Clearing and grubbing direct costs	$ 75,000
Total direct costs	3,053,000
Total job indirects and markup	670,000

$$\text{Distributed amount} = \frac{75,000}{3,053,000} \times 670,000 = \$ \ 16,460$$

Item	Direct cost	Distributed cost	Total cost	Unit cost*
Clearing and grubbing	75,000	16,460	91,460	1,829.20
Rock excavation	622,000	136,500	758,500	15.17
Common excavation	210,000	46,090	256,090	1.28
Select fill	571,000	125,310	696,310	2.32
Paving	1,500,000	329,180	1,829,180	18.29
Structures	35,000	7,680	42,680	42,680.00
Landscaping	40,000	8,780	48,780	48,780.00
Totals	3,053,000	670,000	3,723,000	

*Found by dividing total cost by number of units

Fig. 18-2   Distribution of other costs.

Total distribution of job indirects and markup will produce the arrangement of costs shown in Fig. 18-2 by total amount and by unit cost. Note that the total estimated cost of the project remains as before. The data from Fig. 18-2 are transferred to the Bid Form to provide the final balanced bid as shown in Fig. 18-3. The extended total cost varies slightly from that of Fig. 18-2 because the Bid Form extensions are made from the unit prices which were rounded off to the nearest penny. The client receiving the bids from several bidding contractors will evaluate the bids on the basis of the extended total, not the unit prices, and the contract will be awarded to the contractor with the lowest extended total price assuming the bid is responsive (i.e., fully complies with bidding instructions). In some cases, particularly on public works contracts, unit prices may be subject to scrutiny if specific contract regulations of the client restrict unbalanced bidding (see the next section).

Item	Unit	Engineer's estimate	Bid price	
			Unit price	Extended cost
Clearing and grubbing	Acre	50	1,829 20	91,460 00
Rock excavation	CY	50,000	15 17	758,500 00
Common excavation	CY	200,000	1 28	256,000 00
Select fill	CY	300,000	2 32	696,000 00
Paving	SY	100,000	18 29	1,829,000 00
Structures	LS	1	42,680 00	42,680 00
Landscaping	LS	1	48,780 00	48,780 00
			TOTAL BID	3,722,420 00

Fig. 18-3   Balanced bid.

UNBALANCING THE BID

The bid developed in the previous section is a *balanced bid* in that all work packages were associated with the most logical of the seven Bid Form work items and job indirects and markup were distributed proportionally. This is not a required procedure. The contractor can choose to unbalance his or her bid by redistributing costs in some other fashion among the seven work items without changing the overall total bid price.

You will recall from Chap. 4 that a unit-price contract is awarded on the basis of extended total cost but paid on the basis of unit prices and actual measured quantities during construction. This feature makes unbalancing of unit-price bids attractive for several reasons.

Unbalancing to Obtain Greater Early Income

The contractor will be paid progress payments as work items are completed. The amount of these payments will be related to the unit prices on the bid. The first work item on this project is clearing and grubbing which must be completed ahead of other work. The contractor's balanced bid for this work item is $1829.20 per acre. If the contractor wants earlier receipt of cash, costs can be moved from later line items to clearing and grubbing. For example, the contractor may choose to move $100,000 from paving to clearing and grubbing. This will change the bid to that shown in Fig. 18-4. Note that only unit prices have changed, not the total bid price on which contract award is based.

Unbalancing for Convenience

The excavation portion of the contract includes both common and rock excavation. Sometimes, a contractor may want to avoid on-site

Item	Unit	Engineer's estimate	Bid price	
			Unit price	Extended cost
Clearing and grubbing	Acre	50	3,829 20	191,460 00
Rock excavation	CY	50,000	15 17	758,500 00
Common excavation	CY	200,000	1 28	256,000 00
Select fill	CY	300,000	2 32	696,000 00
Paving	SY	100,000	17 29	1,729,000 00
Structures	LS	1	42,680 00	42,680 00
Landscaping	LS	1	48,780 00	48,780 00
			TOTAL BID	3,722,420 00

Fig. 18-4 Unbalancing for early income.

Item	Unit	Engineer's estimate	Bid price	
			Unit price	Extended cost
Clearing and grubbing	Acre	50	1,829 20	91,460 00
Rock excavation	CY	50,000	4 06	203,000 00
Common excavation	CY	200,000	4 06	812,000 00
Select fill	CY	300,000	2 32	696,000 00
Paving	SY	100,000	18 29	1,829,000 00
Structures	LS	1	42,680 00	42,680 00
Landscaping	LS	1	48,780 00	48,780 00
			TOTAL BID	3,722,920 00

Fig. 18-5   Unbalancing for convenience.

arguments over excavation classification and so will treat all excavation alike for bid and payment purposes. This is done by separately costing rock, common, and other categories of excavation but totaling their costs together and then dividing by the sum of their volumes to produce a single unit price for all classifications of excavation. For our example, the resultant bid would be that of Fig. 18-5. Should this bid be successful, the contractor will be paid $4.06 for each CY of excavation whether common or rock.

Unbalancing for Greater Profit

If a contractor has reason to believe that actual quantities will be significantly different than estimated quantities, this can be put to his or her advantage by loading expected overrun items and underbidding the underrun items. For our example assume that the contractor's field investigation of the project causes him or her to estimate that actual quantities of excavation are

Rock excavation	10,000 CY (down 40,000 CY)
Common excavation	240,000 CY (up 40,000 CY)

Without changing the total bid, the contractor reallocates the extended costs of these two work items to load the common excavation. The extreme condition would be to allocate all costs to common excavation. From the balanced bid (Fig. 18-3), the extended costs of the two categories total $1,014,500. If this is allocated 100% to common excavation, the bid will be as shown in Fig. 18-6. If the contractor has guessed correctly, the effect will be to increase income from the project to the

Items	Unit	Engineer's estimate	Bid price	
			Unit price	Extended cost
Clearing and grubbing	Acre	50	1,829 20	91,460 00
Rock excavation	CY	50,000	0 00	0 00
Common excavation	CY	200,000	5 07	1,014,000 00
Select fill	CY	300,000	2 32	696,000 00
Paving	SY	100,000	18 29	1,829,000 00
Structures	LS	1	42,680 00	42,680 00
Landscaping	LS	1	48,780 00	48,780 00
			TOTAL BID	3,721,920 00

Fig. 18-6   Unbalancing for greater profit.

Item	Unit	Actual quantity	Amount paid	
			Unit price	Extended cost
Clearing and grubbing	Acre	50	1,829 20	91,460 00
Rock excavation	CY	10,000	0 00	0 00
Common excavation	CY	240,000	5 07	1,216,800 00
Select fill	CY	300,000	2 32	696,000 00
Paving	SY	100,000	18 29	1,829,000 00
Structures	LS	1	42,680 00	42,680 00
Landscaping	LS	1	48,780 00	48,780 00
			TOTAL PAID	3,924,720 00

Extra profit $ 202,800 00

Fig. 18-7   Amounts paid on a bid unbalanced for profit.

amount shown in Fig. 18-7. Obviously, if the contractor has guessed incorrectly and there is more rock excavation and less common excavation than estimated, the contractor will suffer an income loss.

Reducing Total Bid to Account for Lower Actual Quantities

Another option available to a contractor who is confident that actual quantities vary significantly from estimated quantities is to use this knowledge to reduce the total bid for the purpose of better ensuring that the contractor's bid is the lowest bid. If the rock or common excavation actual quantities are 10,000 CY and 240,000 CY as mentioned above, the true costs of these work items will be

From Fig. 18-2:

$$\text{Rock direct cost/CY} = \frac{\$622,000}{50,000\text{ CY}} = \$12.44/\text{CY}$$

$$\text{Common direct cost/CY} = \frac{\$210,000}{200,000\text{ CY}} = \$1.05/\text{CY}$$

Adjusted direct cost for new quantities:

Rock cost = \$12.44 × 10,000 CY = \$124,400

Common cost = \$1.05 × 240,000 CY = <u>252,000</u>

$\qquad\qquad\qquad\qquad\qquad\qquad\qquad\quad$ \$376,400

Distributed cost (assume no change):

Rock excavation . . . . = $\quad$ 136,500

Common excavation . . . . = <u>$\quad$ 46,090</u>

$\qquad\qquad\qquad\qquad\qquad$ TOTAL $\qquad$ \$558,990

The contractor can then choose to distribute the entire cost to common excavation which, in this case, would appear as

$$\text{Bid price} = \frac{\$558,990}{200,000\text{ CY}} = \$2.79$$

The actual bid will be as shown in Fig. 18-8 and the payment, if he or she is correct in his or her estimates of quantities, will be shown in Fig.

Item	Unit	Engineer's estimate	Bid price	
			Unit price	Extended cost
Clearing and grubbing	Acre	50	1,829 20	91,460 00
Rock excavation	CY	50,000	0 00	0 00
Common excavation	CY	200,000	2 79	558,000 00
Select fill	CY	300,000	2 32	696,000 00
Paving	SY	100,000	18 29	1,829,000 00
Structures	LS	1	42,680 00	42,680 00
Landscaping	LS	1	48,780 00	48,780 00
			TOTAL BID	3,265,920 00

Fig. 18-8   Lowering a bid to account for lowered quantities.

Item	Unit	Actual quantity	Amount paid	
			Unit price	Extended cost
Clearing and grubbing	Acre	50	1,829 ¦ 20	91,460 ¦ 00
Rock excavation	CY	10,000	0 ¦ 00	0 ¦ 00
Common excavation	CY	240,000	2 ¦ 79	670,788 ¦ 00
Select fill	CY	300,000	2 ¦ 32	696,000 ¦ 00
Paving	SY	100,000	18 ¦ 29	1,829,000 ¦ 00
Structures	LS	1	42,680 ¦ 00	42,680 ¦ 00
Landscaping	LS	1	48,780 ¦ 00	48,780 ¦ 00
			TOTAL PAID	3,378,708 ¦ 00

Fig. 18-9   Amounts paid on a lowered bid.

18-9. This approach has given the contractor a very competitive bid while still returning a bonus profit above costs. It should be noted that the calculations used for this example kept the distribution of job indirects and markup costs as in the earlier balanced bid.

Example of Bid Schedule

Figure 18-10 is a summary of the two lowest bids received on an actual project. Bidder 1 was the successful bidder with the lowest extended total costs. Review of the unit prices for the two bids shows evidence of apparent bid unbalancing. Note that both bidders discounted rock excavation almost completely. Bidder 1 also discounted boulder excavation, but Bidder 2 bid $10.00/CY. Bidder 1 bid exactly twice as much for bank-run gravel as Bidder 2. Both bidders discounted lumber so apparently added that cost elsewhere. On clearing and grubbing, Bidder 1 has apparently loaded the item for early cash while Bidder 2 bid a nominal amount.

Accepting Unbalanced Bids

A certain amount of unbalancing of bids is accepted practice but clients reserve the right to reject a bid as being unresponsive if they consider the unbalancing to be extreme. An extreme example of an unbalanced bid is one which puts the total project cost in the work item to be completed first, all other items being bid as zero cost. Obviously, a client would consider this unresponsive because there is no incentive to complete the work, even if a percentage is retained, after the first item is complete. In some cases a client will specify a maximum value or

		Unit Prices	
		Bid 1	Bid 2
RCP, 66 in., cl 4 .... LF	1,100	177.00	175.00
60 in. .... LF	1,570	160.00	160.00
60 in., cl 5 .... LF	1,010	180.00	180.00
54 in., cl 4 .... LF	1,245	132.00	145.00
48 in. .... LF	3,660	76.00	65.00
18 in. .... LF	2,200	30.00	28.50
VCP, 8 in. .... LF	765	20.00	18.00
6 in. .... LF	240	30.00	20.00
Rock excavation .... CY	4,500	0.01	0.01
Boulder excavation.......... CY	500	0.01	10.00
Ty AMH base .... EA	16	2000.00	1200.00
48 in. diameter riser .... VF	260	80.00	50.00
Ty DMH .... VF	160	100.00	75.00
Special structure 1 .... LS	1	14000.00	8000.00
2 .... LS	1	13000.00	8000.00
3 .... LS	1	12000.00	12000.00
4 .... LS	1	13000.00	10000.00
Diversion structure .... LS	1	20000.00	15000.00
Bank-run gravel .... CY	6,000	4.00	2.00
Temporary paving, $1\frac{1}{2}$ in. ...... SY	16,000	2.00	2.75
3 in. ...... SY	200	4.50	5.00
Permanent paving, 3 in. ...... SY	2,500	4.50	5.00
On 6 in. concrete slab ...... SY	8,000	10.50	11.50
$1\frac{1}{2}$ in. full width .... SY	7,500	2.00	3.00
Lumber .... MBF	500	0.01	0.00
RCP drains, 15 in. .... LF	1,750	18.00	11.00
Clearing and grubbing .... LS	1	26000.00	3000.00
Dewatering and drainage ..... LS	1	12000.00	15000.00

Totals: Bid 1 ... $1,615,365
        Bid 2 ... $1,649,785

Fig. 18-10 Bid schedule for a sewer project.

range of values for key work items to prevent extreme unbalancing. A phrase that has appeared in some government contracts is

A bid which does not contain separate bid prices for items identified as subject to a cost limitation may be considered unresponsive. A bidder by signing his bid certifies that each price bid on items subject to a cost limitation includes an appropriate apportionment of all applicable estimated costs, direct and indirect, as well as overhead and profit. Bids may be rejected which have been materially unbalanced or exceed cost limitations.

SUMMARY

Developing the cost estimates for unit-price contracts is basically the same as for lump-sum contracts. The project is planned and then estimated by work packages. Translating work package costs into a final bid does differ for unit-price contracts as compared to lump-sum contracts. The nature of the unit price bid gives a contractor considerable flexibility to reallocate actual costs among Bid Form work items to alter cash flow, simplify field measurement, or to maximize profit should the contractor win the contract.

REVIEW

Exercises

1. Assuming that actual quantities of rock excavation and common excavation were 10,000 CY and 240,000 CY, respectively, in the example problem of the text, determine the amount paid the contractor if he or she had unbalanced the bid for convenience as in Fig. 18-5. Assuming that costs for each item of work are as given in Fig. 18-2, did the contractor make or lose money?

2. A Bid Form for a project contains these work items and estimated quantities

Work Item	Unit	Estimated Quantity
Common excavation	CY	500,000
Rock excavation	CY	200,000
Structures	LS	—

Your collection sheet calculations indicate that your direct costs for common excavation are $0.95/CY, rock excavation costs $12.00/CY, structures will cost $400,000 and total job indirects and markup should be $140,000.

a. Develop a balanced bid.
b. Unbalance the bid for convenience in costing excavation by equalizing their unit costs.
c. If you have reason to think that rock excavation quantities are considerably less than estimated on the Bid Form while common excavation is higher, develop an unbalanced bid to maximize profit.

Questions in Review

1. What determines the successful bidder on a unit-price contract?
2. What determines the amount paid a contractor on a unit-price contract?
3. What is bid unbalancing? Why is it done?

Discussion Questions

1. In this chapter a phrase designed to minimize bid unbalancing was quoted from a federal government bidding document. The federal government has the power to enforce this requirement since a contractor's books are subject to audit by the General Accounting Office. If you were a private client and included such a statement in your Instructions to Bidders, how would you enforce it?

2. You are a private client intending to solicit bids on a unit-price contract. How would you write your Instructions to Bidders or what other actions would you take to minimize the possibility of serious cost overruns or other problems due to bid unbalancing?

19 APPROXIMATE ESTIMATING

INTRODUCTION

Most of the material contained in preceding chapters has been directed toward the ultimate preparation of a definitive estimate: an accurate estimate based upon well-defined plans and specifications. Mention has been made of those occasions when the plans and specifications are not totally available or the project is defined only conceptually. In such cases, the estimate must be based upon prior experience with similar work and/or the judgment of experienced estimators and construction planners. Such estimates are approximate. Approximate estimating is used in the planning of every project. Very early in a project's planning stage the clients need approximate cost estimates for all options under consideration so that they can choose the plan to be followed and initiate budgetary actions. As the design progresses, updated approximate estimates will be expected since the clients reserve the right to modify or cancel the project should estimates prove unfavorable. Of course, as the design becomes more complete, the accuracy of the approximate estimates approach the accuracy of a definitive estimate. This chapter will describe some of the methods by which approximate estimates can be developed.

BASE UNIT ESTIMATES

At the very earliest stage of its planning, a project is defined only by its intended output or general dimensions. Estimating costs at this point can be done by using data from similar types of projects which are expressed in terms of cost per base unit. Typical base units are

Buildings. Square foot of floor area or cubic foot of volume. May be divided into categories such as single story or multiple story, heated and unheated space, framing types, etc.

Process Plants. Unit of output. Usually, costs will be expressed for different ranges of plant capacity since unit costs go down as size increases.

Power Plants. Megawatts of electrical capacity. Would be broken down by nuclear, coal-fired, gas-fired, and capacity range of each.

Waste or Water Treatment Plants. Gallons of waste or water treated daily. Would be different capacity ranges as for process plants.

Highways. Mile. Would be related to terrain, soil types, location, number of lanes, and wearing surface.

One of the most commonly used base units is that for residential housing: square feet of space. Every housing contractor, mortgage banker, insurance company, and realtor can estimate quite accurately the value of a given design using square foot factors appropriate to the unit being evaluated.

Costs per base unit are area sensitive since construction costs vary so widely for reasons mentioned throughout this book. Also, like all costs, they are time sensitive due to the effects of inflation.

Base unit costs are available from several sources. Contractors specializing in certain types of work will accumulate such data for their work. Engineering sections of industries will maintain statistics which permit development of base unit costs relating to their industry. Commercial cost sources (see later discussion) provide this information on many categories of projects.

SCALE OF OPERATIONS ESTIMATES

Another form of approximate estimate appropriate for the first stage of planning of process plants relates costs to output according to the formula

$$C_B = C_A \left(\frac{Q_B}{Q_A}\right)^x$$

where: C_A is the cost of an existing plant producing the same product
       C_B is the cost of the proposed plant (to be determined)
       Q_A is the capacity of the existing plant
       Q_B is the capacity of the proposed plant
       x is an exponent (usually 0.65 or less)

The appropriate exponent would be established by the industry
involved based upon its past experience. The formula estimates the
cost of the proposed facility (Plant B) in the year that the reference
facility (Plant A) was built. Therefore, another step in the estimating
process will be the accounting of inflationary cost increases projected
to the midpoint of construction of the proposed facility (see Chap. 3).

PARAMETER ESTIMATES

As the design of a project progresses through the preliminary design
phase, the project's key features (foundation, framework, architectural
features and general dimensions, and major installed equipment items)
are generally defined although not necessarily fully designed and exactly
dimensioned. At this point, an approximate estimate can be made
using parameters derived from similar projects. This method will be
explained using the parameter system of *Engineering News-Record* magazine for buildings. Each quarter, this magazine publishes data from a
group of recently completed building projects in the United States from
which parameter data have been derived. This service has been provided
for a number of years and complete sets of past parameter reports may
be purchased by writing the Construction Economics Department, *Engineering News-Record*, 1221 Avenue of the Americas, New York, N.Y.
10020. Figure 19-1 shows typical data for a warehouse/office building
built in San Francisco in 1976.

The project is described in the first several lines of data. Following that are listed the standard 14 parameter measures used in all reports.
Opposite each parameter measure is the quantity for this project, if applicable. For the warehouse/office described in Fig. 19-1, parameter
measures 3 and 11 are not applicable. The next two sections, Other
Measures and Design Ratios, give other descriptive data for the project.
Other Measures also serve as parameters for certain of the trades as will
be discussed below. Following these sections are listed building components by trade. This listing is standardized so only those trades applicable will have entries. For each line of data, information on parameter cost and total cost is provided. Under Parameter Cost the first
subsection is "Code." The code number entered refers to one of the
standard 14 parameter measures listed or OM is given, meaning Other
Measures. This parameter measure is considered to be the one of greatest significance costwise with respect to the trade being measured. For
example, Structural Steel shows a code 2 meaning that structural steel
costs are most closely related to parameter 2, Gross Area Supported.
The unit of measure is that of the parameter, in this case SF for square
feet. The parameter cost is $4.78/SF of gross-supported area. The
Total Cost columns give the total cost of each trade and the percentage

Type of Building	Warehouse/office
Location	San Francisco
Construction start/complete	Feb. '76/Oct. '76
Type of owner	Private
Frame	Struct. Steel
Exterior walls	Curtain wall metal/glass
Special site work	—
Fire rating	Type 5

PARAMETER MEASURES:

1. Gross enclosed floor area	15,000 sf
2. Gross area supported (excl. slab on grade)	18,000 sf
3. Total basement floor area	—
4. Roof area	6,000 sf
5. Net finished area	11,000 sf
6. No. of floors including basements	3
7. No. of floors excluding basements	3
8. Area of face brick	5,500 sf
9. Area of other exterior wall	11,000 sf
10. Area of curtain wall incl. glass	1,600 sf
11. Store front perimeter	—
12. Interior partitions	1,070 lf
13. HVAC	35 tons
14. Parking area	900 sf

OTHER MEASURES:

Area of typical floor	6,000 sf
Story height, typical floor	12 ft
Lobby area	100 sf
No. of plumbing fixtures	16
No. of elevators	1

DESIGN RATIOS:

A/C ton per building sq ft	0.0023
Parking sq ft per building sq ft	0.06
Plumbing fixtures as per building sq.ft.	0.0011

TRADE	Code	Parameter Cost: Unit	Parameter Cost: Cost	Total Cost: Amount	Total Cost: %
General conditions and fee	1	sf	5.00	75,000	10.20
Sitework (clearing and grubbing)	1	sf	2.33	35,000	4.76
Utilities	—			—	—
Roads and walks	1	sf	0.25	3,800	0.52
Landscaping	5	sf	0.02	200	0.03
Excavation	2	sf	0.22	4,000	0.54
Foundation	2	sf	2.33	42,000	5.71
Caissons and pilings	2	sf	2.22	40,000	5.44
Formed concrete	2	sf	0.67	12,000	1.63
Exterior masonry	5	sf	3.64	40,000	5.44
Interior masonry	—			—	—
Stone, granite, marble	5	sf	0.09	1,000	0.14
Structural steel	2	sf	4.78	86,000	11.70
Misc. metal, incl. stairs	(included in struct. steel)				
Ornamental metal	—			—	—
Carpentry	5	sf	5.90	64,850	8.82
A/C enclosures	—			—	—
Waterproofing and dampproofing	4	sf	0.08	500	0.07
Roofing and flashing	4	sf	1.50	9,000	1.22
Metal doors and frames	5	sf	0.45	5,000	0.68
Metal windows	5	sf	2.27	25,000	3.40
Wood doors, windows, and trim	5	sf	1.37	15,100	2.05
Hardware	5	sf	0.29	3,200	0.44
Glass and glazing	(included in windows)				
Store front and lobby	(included in windows)				
Curtain wall	(included in windows)				
Lath & plaster	(included in drywall)				
Drywall	12	lf	42.06	45,000	6.12
Tile work	5	sf	0.55	6,000	0.82
Terrazzo	—			—	—
Acoustical ceiling	5	sf	0.55	6,000	0.82
Resilient flooring	5	sf	0.03	300	0.04
Carpet	5	sf	1.09	12,000	1.63
Painting	5	sf	0.71	7,800	1.06
Toilet partitions	12	lf	2.85	3,050	0.41
Special waste treatment	—			—	—
Venetian blinds	—			—	—
Special equipment	5	sf	0.23	2,500	0.34
Elevators	om	ea	21,400.00	21,400	2.91
Plumbing	om	ea	1,125.00	18,000	2.45
Sprinklers	5	sf	1.27	14,000	1.90
HVAC	13	tons	2,177.14	76,200	10.37
Electrical: Contracts				—	—
Fixtures	1	sf	2.45	36,800	5.01
Miscellaneous trades	1	sf	1.42	21,300	2.90
Parking: Outside, enclosed	—			—	—
Open, paved	14	lf	3.33	3,000	0.41
TOTAL	1	sf	49.00	735,000	100

Fig. 19-1 *Engineering News-Record* parameter data. (Reprinted from *Engineering News-Record*, June 23, 1977, copyright McGraw-Hill, Inc., all rights reserved.)

of the total construction cost represented by this trade. For structural
steel the total cost was $86,000 and this represented 11.7% of the total
cost.

The 14 parameter measures and other measures listed in Fig. 19-1
will be fairly well defined during the preliminary design stage of a building. Thus, if an estimator has available sets of parameter data on similar
buildings, he or she can develop a reasonable approximate estimate at
the preliminary design stage. As always, the data are area and time
sensitive so appropriate adjustments must be made to the raw data.

The example parameter system is for buildings but the same technique can be applied to any category of construction.

FACTOR ESTIMATING

Factor estimating is another technique applicable to the early design
stages of a structure. It is particularly applicable to industrial construction. During the preliminary design stage a project's major structures and components are defined in general terms such as approximate
area, volume, capacity, or type. To factor estimate, the project is
treated as a whole or divided into its several components (structures or
systems). For each component, major items of installed equipment are
extracted and their costs estimated individually. Other quantities are
factored using charts or data appropriate to the structure. For example,
on a nuclear power plant estimate, one major structure is the reactor
containment. It will contain the reactor itself, steam generators (if
applicable), pumps, cranes, and numerous other items of installed
equipment typically found in such a structure. Each of these can be
itemized and their cost reasonably approximated. The civil, electrical,
and mechanical construction quantities are then estimated using factor
charts derived from previous experience on such structures. For example, cubic yards of concrete, tons of reinforcing steel, numbers of
cadwelds, linear feet of piping, linear feet of electrical cable, and other
quantities can be related to the reactor size. Construction equipment-
hours and employee-hours are estimated the same way. These labor,
material, installed equipment, and construction equipment quantities
are converted to current or projected costs, using current cost figures
for the units involved and combined with costs of the other structures
and systems in the plant to provide the total direct cost estimate. Engineering employee-hours, construction management employee-hours,
construction plant, and other job indirect costs can be similarly estimated using curves developed from prior experience.

Figure 19-2 is a form of graph for use with factor estimating. This
graph is for a hypothetical type of structure and the quantity being

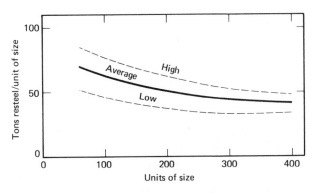

Fig. 19-2   Graph for factor estimating.

measured is tons of resteel per unit of size. The graph contains high and low curves as well as an average curve, each having been plotted on the basis of experience data. For a 200-unit plant the average amount of resteel is about 51 tons/unit for a total of 10,200 tons for the total structure. If some feature of the proposed structure might cause the estimator to use a higher or lower figure, the range is available for such adjustments.

Contractors will develop their own factor charts. A contractor specializing in residential construction can develop factor charts for materials and employee-hours used in home construction with quantity requirements plotted against square or cubic footage of residence. A major advantage of factor charts as compared to base unit or parameter cost techniques is that data can be maintained in terms of quantities of resources, not their cost, so the information is less perishable.

COMMERCIAL COST REFERENCE PUBLICATIONS

Commercial cost reference publications, most of which are revised at least annually, are available to facilitate both approximate and detailed estimating. The R.S. Means Co., Inc., 130 Construction Plaza, Duxbury, Massachusetts 02332 publishes several estimating guides. Their *Building Construction Cost Data* book is printed annually and is composed of several components. The major section of the book is a cataloging of material and installation costs for virtually all types of work that can be included in the 16 Uniform Construction Index divisions for building construction. The data presented are average for the United States and assume union wage scale. Also included are ranges of cost per square foot for all common types of buildings. A special section presents cost indices for conversion of average data to specific areas of the United States. A companion publication to this manual is the *Building*

Systems Cost Guide which is designed to assist planners in the conceptual stage of design by presenting costs for alternate building systems and components. Other Means publications cover labor rates, cost indices, and mechanical and electrical cost data.

The *Dodge Digest of Building Costs and Specifications,* published by the McGraw-Hill Information Systems Company, 1221 Avenue of the Americas, New York, New York 10020, catalogs actual projects by building type and projects are described as to design features and costs. It is a subscription publication and is updated semiannually. Included in the publication are cost indices for many areas of the United States to permit adaptation of information to locations other than those reported. Dodge also publishes a *Dodge Manual for Building Construction* similar to the Means manual, the *Dodge Construction Systems Costs* manual which provides cost data for functional components of a building, the *Dodge Guide to Public Works and Heavy Construction Costs* which features information on nonvertical construction, and several other specialty publications suitable for approximate estimating.

Richardson Engineering Services, Inc., P.O. Box YE, Solana Beach, California 92075 publishes two sets of estimating standards, one for general construction and one for process plants. Data in these volumes is suitable for use in both approximate and definitive cost estimating.

The *National Construction Estimator,* published by the Craftsman Book Company, 542 Stevens Avenue, Solana Beach, California 92075, is one of a family of estimating guides covering both construction and remodeling of buildings.

Publications of the type listed have proven invaluable to contractors whose own data base is incomplete. However, as emphasized throughout this text, the most accurate data for a contractor are those compiled through his or her own experience.

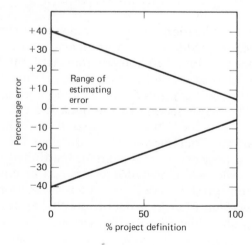

Fig. 19-3   Estimating accuracy versus project definition.

APPROXIMATE ESTIMATING ACCURACY

In Chap. 1 the many problems associated with the Montreal Olympics complex were discussed. The overall effect was to drive actual costs of construction more than 1000% over the preliminary estimated costs. While such overruns are not the rule, one must expect preliminary estimates to be subject to considerable error and for the level of accuracy to improve as the project becomes more refined. Figure 19-3 is an approximation of range of estimating error that can be expected with various degrees of project definition.

SUMMARY

Approximate estimates are required during the preliminary planning stage and throughout the design stages of a project. Various methods for developing these estimates are in use, some based upon contractor or client historical experience, others developed by commercial organizations providing such specialized service.

The level of estimating accuracy is obviously a function of the level of project definition. Estimates prepared during the early planning stages can be expected to vary 40% or more from actual costs as compared to definitive estimates which should be capable of zeroing in on costs within 5%.

REVIEW

Exercises

1. Assume it is 1978. A four-story warehouse/office building is in the conceptual design stage for Houston, Texas. If built, construction will occur during 1979. Approximate dimensional information derived from the preliminary drawings are

    a. Gross enclosed floor area   18,000 SF
    b. Gross area supported   20,000 SF
    c. No basement
    d. Roof area   6,000 SF
    e. Net finished area   13,000 SF
    f. Number of floors   4
    g. Area of exterior wall, total   17,000 SF
           Glass   14,000 SF
           Masonry   3,000 SF
    h. Interior partitions   1,300 LF
    i. HVAC   60 tons
    j. Parking area   1,200 SF
    k. Elevators   2
    l. Plumbing fixtures   20

Using the parameter cost information of Fig. 19-1, estimate the cost of this structure. For this purpose assume that the construction cost index for buildings in 1976 was 1425 and is projected to be 1700 in 1979 on a national average. Also assume that the relative cost indices between San Francisco and Houston are in the ratio of 118 to 100.

2. A process plant built in 1974 has a capacity of 4 million units per year. A new and similar process plant to produce 8 million units per year is being considered for construction in 1980 in a nearby location. The cost of the 1974 plant was $38 million. What do you project the cost of proposed plant will be? The cost index for 1974 was 1200 and is projected to be 1800 in 1980.

Questions in Review

1. What is the difference between base unit and parameter estimating?
2. What is the major advantage of factor charts over parameter costs?
3. At what stage in planning can parameter costs be used?

Discussion Questions

1. Although most clients do not expect contractors to bid a fixed price on a contract with incomplete designs, there are clients who do. This forces contractors to utilize factors, parameters, or other approximate estimating techniques to prepare most of the estimate. Can you think of any ways the contractor can protect himself from overruns in such a situation other than by adding a significant contingency percentage?

2. Most commercial cost guides are derived from voluntary input from contractors. If you were a contractor and had just made a killing on a fixed-price contract, would you be willing to provide your true costs to one of these services?

INDEX

INDEX

Descriptive specification, 60-61
Design:
drawings, 62
effect on productivity, 2-4, 6
130-31
Design-build, 10, 12
Dewatering, 256
Ditching (trenching) machines, 154
Divisions of specifications, 60-61
Dodge Construction Systems Costs, 318
Dodge Digest of Building Costs and Specifications, 318
Dodge Guide to Public Works and Heavy Construction Costs, 318
Dodge Manual for Building Construction, 318
Double-breasted operations, 102
Dragline, 153
Drawings, 6, 62
Dry walls, 256

E

Earned employee-hours, 89-91, 297-98
Earthwork:
collection sheet preparation, 220-22
end area measurement method, 215-16
equipment (*See* Construction equipment types)
grid measurement method, 216-19
measurement, 201-2, 213-20
Economic equipment life, 41, 167, 182-83
Electrical work:
buildings, 256
cable tray installation, 250-51
collection sheet preparation, 249-52
grounding, 248
installation sequence, 248-49
underground ducts, 249-50
wire pulling & terminations, 249, 251-52
Elevating scraper, 147
Elevators and escalators, 256
Employee-hours:
actual, 90-91, 297

Employee-hours (*continued*):
budgeted, 89-91, 296
earned, 89-91, 297-98
estimated, 89-91, 296
scheduled, 89-91, 298
target, 89-91, 297-98
Engineer, 10
Engineering:
control of costs, 295-98
cost estimation, 294-95, 298-99
work packaging, 293-94
Engineering-construction firm, 10
Engineering News-Record, 123, 181, 314
Equal Employment Opportunity Act of 1972, 95
Equipment (*See* Administrative equipment; Construction equipment; Installed equipment; Mechanical equipment; Production equipment; Support equipment)
Equipment division in a company, 16, 183
Escalation clauses, 56, 210
Estimating (*See* Bid development)
Excavators, 153
Exclusive recognition, 99-100
Experience rating:
Unemployment Insurance, 109
Worker's Compensation, 110-11

F

Federal civil works construction, 14
Federal Mediation and Conciliation Service, 95
FICA (Social Security), 108-9
Finance staff sections, 17-18
Finishing concrete, 230, 232
Fixed-price contracts, 52-53
Flooring, 256
Force account labor, 11
Forklifts, 158
Forming:
collection sheet preparation, 224, 225
types, 223
work measurement, 223-24
Front loader, 149
Fuel costs, 171
Fuel trucks, 160